Basics of
Modern Cosmology

A. D. Dolgov
M. V. Sazhin
Ya. B. Zeldovich

Editions Frontières

Translated from the Russian by :

Tomasz Bulik and Michał Bulik

ISBN   2-86332-073-4 (hard cover)
        2-86332-084-X (soft cover)

ISSN   0982-5657

Editions Frontières
B. P. 33
91192 Gif-sur-Yvette Cedex - France

Printed in Singapore by Fong and Sons Printers Pte. Ltd.

EDITOR'S PREFACE

During the last decades, modern physics has undergone a series of bifurcations and unifications. From possible symmetry structures of particles have emerged grand unification symmetries; from the tentative idea of supersymmetry we have entered the domain of string and superstring field theories; from being the least firmly rooted of physical theories, cosmology has formed a new partnership with particle physics and the results have exceeded everyone's expectations.

In such rapidly developing disciplines it is often difficult to educate oneself in the latest ideas. Summer School proceedings play an important role in surveying new fields but often lack the detailed pedagogy required of a textbook. There is a real need for books from which the reader with an ordinary background in general physics can learn to handle the new ideas and follow current work.

The *"Basics"* series of books aims to provide timely introductions to the most important areas of fundamental physics. We at Editions Frontières hope that these books will help to clarify and unify contemporary developments, bringing order and understanding to specialists and non-specialists alike.

It is a pleasure to welcome A. D. Dolgov, M. V. Sazhin and Ya. B. Zeldovich to this series of *"Basics"* books. "Basics of modern Cosmology" is an updated version of "Cosmology of the early Universe" published by the Moscow University Press (Izdatelstvo Moskovskovo Universiteta) in 1988. It provides a thorough description of the deep interaction between elementary particle physics and cosmology, leading to an understanding of the early stages of the Universe. It will be read with pleasure and profit by scientists and educators, by students as well as by specialists.

J. Trân Thanh Vân
December 1990

Contents

Foreword

Modern cosmology is developing mainly in two directions. The first one consists of investigation of the structure of the Universe, background radiation and direct searches for the hidden mass. New theoretical ideas, as well as observational data (especially those obtained during the past 10 years) are used for this purpose.

The other direction in cosmology is concerned with revealing early periods in the history of our Universe. This domain of cosmology is popular not only among cosmologists but also physicists especially those dealing with elementary particle problems and fundamental interactions.

The early stages of the evolution of the Universe provide a natural laboratory in which extreme conditions were realized. It is impossible even with modern technology to realize an experiment that would involve such high density and temperature as those which are revealed by cosmology.

For this reason it is wrong to think that the theory of the Early Universe is developed on an experimental basis. The way we view this problem has changed - today we look for proof of the elementary particle theories and Unification Models in cosmology. Sometimes the Early Universe is called "the poor man's accelerator", meaning that cosmological studies can give hints on particle interactions at such high energies, which are beyond the scope of the present day human possibilities.

A very productive cooperation exists between cosmology on one hand and elementary particle physics and quantum field theory on the other.

The cosmological tests of elementary particle theories at very high energies are a unique way to compare them with physical reality. On the other hand the results of these theories made it possible to answer fundamental questions about the Universe, which were quite recently considered as unanswerable within human capacity. These were the problems of general properties of the Universe like homogeneity, isotropy, the energy density and why the latter is very close to its critical value corresponding to a spatially flat Universe. Fundamental physical laws permit us to understand the origin and spectrum of density fluctuations at early stages, which because of gravitational instability later give rise to galaxies and their clusters. Just a few years ago all these properties and the values of these parameters were taken as initial conditions of cosmological models. All these conditions looked unnatural in the framework of our previous knowledge. Now they all can be obtained as a result of a rather general dynamics which leads to the exponential law of the Universe expansion i.e. to the de Sitter model. This is called the inflationary model of the Universe.

Today we even hope to understand and describe mathematically the creation of Universe and to find physical reasons for its expansion from the initial singularity. According to contemporary theories this process looked like a quantum jump, very similar to tunnelling in α decay, from a quantum spacetime state to the classical one. We are also on the way to discover why our Universe is 4-dimensional. The natural dimension for the quantum field theory is probably D=10. Strictly speaking, this refers to the superstring theory, which hopefully may lead to a self-consistent unified theory of all interactions. Probably this theory will even explain why the remaining six dimensions are compact and are not observed in our laboratories with the available energies. Not all the problems are solved but the perspectives seem exciting.

This book is based on lectures given by Ya.B.Zeldovich at the physical faculty of the Moscow State University. It is written mainly for theoretical and observational astronomers. It will be interesting to other physicists wishing to get acquainted with the modern cosmology. We assume that the reader is familiar with special relativity. The knowledge of general relativity is not necessary but might be useful. We assume that the reader possesses mathematical background on the level of university courses and the basics of quantum mechanics. The knowledge of the fundamentals of particle theory is not assumed. They are briefly discussed in the books by L.B.Okun "α, β, γ ..,Z" (1985) and "Fizika elementarnych czastic" (1984). Those willing to learn the standard cosmology should read "The first three minutes" by S.Weinberg and more detailed books by Ya.B.Zeldovich and I.D.Novikov "Strojenie i ewolucja Vselennoj" (this book has been translated into English) or "Gravitation and Cosmology" by S.Weinberg. These books, however, do not cover the exciting development of cosmology in recent years. We hope that this book will help to fill this gap.

Except for several basic papers we refer only to books and review articles which can help the reader to get a deeper insight into the problems considered. It is especially true for particle physics. All the references that might be needed can be found in the above mentioned books and papers. As far as cosmology is concerned the references are given to recent articles that introduce new ideas about the early Universe.

The readers to whom the particle physics is new might find the first chapters difficult. After reading chapter 4 all those difficulties should hopefully be overcome. Generally, the chapters are independent but we advise to read the book in succession. Throughout this book a system of units is used in which $\hbar = c = k = 1$ (see eg. book [25]) that is unusual to astronomers.

As the subject discussed in this book is developing very quickly we must note that the work on this book was finished in July 1986. Therefore the book does not reflect the recent developments in the field. Chapters 1, 2, 6, 9 were written by M.V.Sazhin, chapters 4, 7, 8 by A.D.Dolgov, chapters 3, 5, 10, 11 together.

Chapter 1

Standard cosmological model

By this term we mean the model of the hot Universe or the Big Bang model which was generally accepted after the discovery of the cosmic microwave background radiation. The dynamics of this model is well known and described by general relativity but the initial conditions leading to the present Universe are mysterious. The recent development in cosmology consists mainly in understanding the physical grounds for these specific initial conditions and does not change the validity of the classical cosmology at later stages.

1.1 Expansion of the universe - the Hubble constant

The modern cosmology is based on the solution of general relativity equations for the isotropic and homogeneous distribution of matter, found 60 years ago by Friedman. The expansion of the Universe predicted by this solution was proved by Hubble (1929) and is given his name. The expansion law is very simple:

$$v = Hr \qquad (1.1)$$

where v is the velocity of an astronomical object at the distance r. The proportionality constant H is called the Hubble constant. Its

9

value is (according to recent measurements):

$$H = (50 - 100)\text{km/s/Mps}$$

The term Hubble constant is misleading. "Constant" means only that the value of H does not depend on distance or direction to the celestial body but it is a function of time. The value $H^{-1} = (20 - 10) \times 10^9$ years equals roughly the age of the Universe from the moment of the singularity. At early stages, the Hubble "constant" was bigger and the velocity of expansion was higher.

According to observations made by Hubble, the value of H was about 500km/s/Mps, which means that the age of the Universe would be a mere 2×10^9 years. Nuclear dating methods and estimation of the age of old star clusters give for the age of the Universe $(20 - 10) \times 10^9$ years. During the fifty years since Hubble's discovery, however, the accuracy of astronomical measurements has improved so that the present value of H is consistent with other estimates of the age of the Universe. The only uncertainty in H is due to difficulties in estimating distances to galaxies. Precision may improve in future years.

Notice that expansion law (1.1) does not choose any particular point in the Universe. A change of reference point together with Galilean transformation leave the Hubble law unchanged. This corresponds to the homogeneity of the Universe.

The expansion of the Universe is not questioned anymore, especially because it is described by a solution of the Einstein equations. The physical grounds for realization of this specific solution were unknown in the standard cosmology. The standard Friedman model could not explain the primary push that caused the expansion of galaxies. The results of particle physics applied to the theory of the early Universe bring us closer to understanding this problem. Going further, we will say that particle physics predicts the state equation with negative pressure, which will be discussed further on. In the relativistic theory of gravity the matter with such an equation of state "antigravitates" and that leads to the expansion of the Universe.

1.2 Background radiation

The most important discovery in cosmology after that made by Hubble was the discovery of electromagnetic background radiation by Penzias and Wilson (1965). The hypothesis about the existence of electromagnetic radiation left over the early hot universe was first made by Gamow in 1946. Strictly speaking, he wanted to describe nucleosynthesis by hot matter processes during the first few minutes. He obtained the relic radiation as a byproduct. Despite his assumptions being not absolutely correct (he thought that all nuclei including heavy ones are produced during the Big Bang [1]) he made a good estimate of the relic radiation temperature (6K). The measurements confirmed the Planck spectrum of the radiation in the wavelength range from tenths of centimeters to millimeters ($\lambda_{max} = 73$cm). The temperature is about 3K in a rather good agreement with previous estimates. The observations indicate high isotropy of the relic radiation. With the exclusion of the simple dipole anisotropy due to the movement of the Earth relative to the background, no other deviations from isotropy at the level of 10^{-5} have been found.

1.3 Matter density in the Universe

The average matter density in the Universe is not so well known as the relic radiation. The dimensionless parameter Ω, which is often used to describe it, is:

$$\Omega = \rho/\rho_c = \rho/(3H^2/8\pi G) \qquad (1.2)$$

where G is the gravitational constant, H is the Hubble constant and ρ_c is the critical matter density. If $\Omega > 1$ ($\rho > \rho_c$) then the Universe is closed and the expansion will change into contraction. This is not necessarily true for the Universe with a non-zero cosmological term (see chapter.5). If $\Omega < 1$ then the Universe is open and will expand

---

[1]It is established now that only light elements have been produced during the first few minutes after the Big Bang (lighter than Li^7). The heavy nuclei have been produced later in stars.

forever. The case $\Omega = 1$ corresponds to a flat Universe with ordinary Euclidean geometry. We mean the geometry of three dimensional space obtained by setting t=const in the four dimensional space-time manifold. This four dimensional manifold of course is not flat. Let us note that there is an arbitrariness in the choice of the time coordinate. We mean here the so called comoving time which is defined by the condition $\rho = const$ (the time axis is perpendicular to the $\rho = const$ hypersurface).

According to astronomical observations the value of Ω is bounded by:

$$0.03 < \Omega < 2$$

The lower bound corresponds to the visible matter. This value is sometimes called Ω_b, the index b meaning that it is ordinary matter made of protons and neutrons (they are generically called baryons).

It is however possible that for some reason a part of the baryonic matter is not visible. For example it might be in the form of ionized gas or dark stars. But even if this were true Ω_b would hardly be greater than 0.2. The arguments based on the primordial nucleosynthesis give Ω_b very close to the lower bound $\Omega = 0.03$ (see section 8).

1.4   Hidden mass

It is generally accepted now that most of the mass in the Universe is in an unknown nonbaryonic form. It is observed only indirectly by its gravitational effects. One can estimate the corresponding value of Ω to be at least about 0.3. This matter forms the hidden mass of the Universe. Two kinds of hidden mass are to be distinguished. The first one makes halo around galaxies and their clusters. The spatial distribution of this form of matter follows that of visible matter. The corresponding Ω is about 0.3. It cannot be excluded that apart from the clustered, inhomogeneous invisible mass there is a uniform background of invisible matter. It is very difficult to see such a background in astronomical observations so the accuracy of these observations is very low. We make the hypothesis about the existence of such a background since the inflationary model, on which the solutions to many fundamental cosmological problems are based, predicts $\Omega_{total} = 1$. Thus Ω corresponding to the background matter is $\Omega_{bg} \approx 0.7$.

It is surprising that at least 90 percent of matter in the Universe is in an unknown form. But this conclusion seems unavoidable. It is very difficult to explain why the bulk of matter, if it is baryonic, escapes observations. There are also independent arguments that all the matter in the Universe cannot be of the baryonic type. The first argument is that the primordial nucleosynthesis gives $\Omega = 0.03$ (see section 8). The second one is based on the theory of the Universe structure formation. It will be discussed in chapter 10. It is important to note that the high isotropy of the background radiation proves that the inhomogeneities of the baryonic matter at early stages must be very small. Hence the galaxies and their clusters could not arise during the available time. The problem can be solved by gravitating matter not interacting with electromagnetic radiation, in other words, by the hidden mass.

1.5   Inhomogeneities

The observations show that the matter in the Universe is distributed very unevenly on small scales. However the Universe is uniform on big scales. The scale characterizing transition from heterogeneity to homogeneity is about 200 Mps. In other words, measuring matter density averaged over a cube with edge smaller than 200 Mps (for example 2 Mps), one obtains a density contrast of order unity. But the density averaged over a cube with edge greater than 200 Mps is homogeneous, that is the value of the density averaged over a large cube does not depend on the position of the cube. Let us notice that the size of the observable part of the Universe is several gigaparsecs so there are about one thousand cubes with 200 Mps edge. This number is large enough to exclude statistical fluctuations and to draw conclusions about the homogeneity of the Universe on large scales.

The measurements of microwave radiation show that in the past the Universe was even more homogeneous . According to recent observations the relic radiation is isotropic with accuracy greater than 3×10^{-5} for angles from $90°$ to $10'$ and 10^{-2} for smaller angles. It follows that degree of homogeneity at the time of hydrogen recombination was about 10^{-4} for the scale which corresponds to the present day distance of 10 Mps and 10^{-1} for 10 ps.

The data on the homogeneity concerning the earlier epoch are not so restrictive. Observations of cosmic deuterium allow us to draw the conclusion that the baryon distribution was rather homogeneous at the time of primordial nucleosynthesis ($t = 100$s), since the deuterium production is very sensitive to baryon concentration.

The inhomogeneities in the early Universe on scales exceeding the horizon size at that time are very much restricted by the data on the large scale anisotropy of the relic radiation, isotropy of the Hubble constant etc.

We can conclude that the early Universe was homogeneous to a high degree on small as well as on large ($R > 1/H$) scales. Hence the assumption of a strictly homogeneous and isotropic Universe is used as a zero-order approximation in the theory of cosmological evolution. Small inhomogeneities, initially present in this smooth background, arose due to gravitational instability. Let us stress that an assumption of primary inhomogeneities is necessary because no large scale structure like galaxies could develop in a strictly homogeneous Universe. Such inhomogeneities have a small influence on early cosmological evolution but later provide seeds for gravitational clustering.

The origin of inhomogeneities and their spectrum was not known even very recently. The hypothesis that they came from quantum or thermal fluctuations gave too small a value for them. The inflationary model has changed the matter completely. It is proved that during the exponential expansion of the Universe quantum fluctuations grow very fast as short wave modes turn into long wave ones. As a result inhomogeneities can even become too large. This problem can be solved by the unnatural smallness of parameters. Let us note that the inflationary model predicts a flat spectrum of fluctuations which more or less agrees with observations.

The theory of gravitational instabilities in Newtonian gravity was developed in the beginning of our century by Jeans (1902). Fourty years later (1946) E.M.Lifschitz generalized it to the Friedman cosmology with relativistic effects taken into account. The inflationary model allows us to find the form of the density fluctuations needed as initial conditions in the theory of gravitational instability. Thus we understand in principle the physical processes leading to the formation of the structure of the Universe but the qualitative final model has not been worked out yet, in particular because little is known about the nature of the hidden mass. Let us recall that $\Omega = \rho/\rho_c$ is close to one, according to modern views, and the baryon component of Ω is only 0.03, i.e. invisible unknown matter forms the dominant part of the mass of the Universe.

1.6 Thermal history of the early universe

Let us now turn to the evolution of the homogeneous Universe at early stages before the development of structure but after inflation. For the time smaller than 10,000 years from the beginning, the dominant part of the energy density was relativistic matter (electromagnetic radiation, neutrinos) for which the equation of state is $p = \rho/3$. A simple approximate formula connecting the age of the Universe and the plasma temperature for this period can be written:

$$T(\text{MeV}) = t^{-1/2}(s) \qquad (1.3)$$

This equation can be obtained by comparing two formulae for the energy density. The first one is the critical density (as we will see later at the radiation dominated epoch $\rho \approx \rho_c$):

$$\rho = \rho_c = \frac{3H^2}{8\pi G} = \frac{3}{32\pi G t^2} \qquad (1.4)$$

where $H = 1/(2t)$ is the Hubble "constant" in this period. The second one is the expression for the energy density of a relativistic gas at temperature T:

$$\rho = \frac{\pi^2}{30}QT^4 \qquad (1.5)$$

where Q is the number of degrees of freedom of different particle species with $m < T$ (e.g. $Q=2$ for photons, as they have two spin states). Each boson degree of freedom adds 1 to Q, each fermion degree of freedom adds 7/8).

Using formula (1.5) we assume that the primeval gas (or maybe it is better to say "plasma") is in state of thermodynamical equilibrium. This is usually true because typically the expansion rate is much smaller than the rate of restoring the equilibrium in the plasma. The closer we are to the beginning, $t \to 0$, the higher the rate of expansion. It is determined by the Hubble constant $H = 1/2t$ (it is assumed that radiation dominates at early stages so $a(t) \sim \sqrt{t}$). However, the particle number density grows even faster, $n \sim T^3 \sim t^{-3/2}$. It ensures that the condition $\sigma n > H$ (where σ is the interaction cross section), which leads to equilibrium at early stages, is satisfied.

At $t \approx 1s$ the primordial plasma temperature was 10^{10}K ≈ 1 Mev. At that time the plasma contained photons, neutrinos and antineutrinos, the above mentioned dark matter (if it does not coincide with massive neutrinos), and electron-positron pairs. If we assume that there were no e^+e^- pairs then at $T \approx 1$ MeV they should appear thanks to the inverse annihilation process : $\gamma + \gamma \to e^+ + e^-$. As the plasma cools down the annihilation process $e^+ + e^- \to \gamma + \gamma$ dominates and the e^+e^- concentration decreases according to the exponential law $n \approx \exp(-m/T)$.

Neutrinos and antineutrinos are in equilibrium with the primordial plasma only for $T > 3 - 5$ MeV ($t \leq 0.1$s) while at smaller temperatures their interactions can be neglected. During the expansion of the Universe they adiabatically loose energy. The present temperature of neutrino - antineutrino pairs is about 2K (if $m = 0$) and their concentration is about $n \approx 75 \, (T/3K)^3 cm^{-3}$.

Protons and neutrons play no role in the energy density at $T \approx 1$ MeV as their number is 10^{-9} to 10^{-10} of the number of light particles (electrons, neutrinos, photons). Closer to the "beginning" at $t < 10^{-2}$ s there were π mesons, K-mesons and proton-antiproton pairs as well as neutron antineutron pairs in the primordial plasma. At $T > 10^{12}$ and respectively $t < 10^{-4}s$ they dissociate producing the so called quark-gluon plasma. Speaking more precisely, as the plasma cools down quarks merge into mesons, protons, neutrons etc. These processes will be discussed in detail further on in this book. For the time being it will only be shown that going backwards to the "beginning", as the temperature grows the primeval plasma should contain all the particles with masses satisfying the condition $mc^2 < kT$.

In particular, when $t \leq 10^{-10}$ or $T \geq 10^{15}$K the intermediate bosons of weak interactions $W^{\pm}$ and Z^0 were abundant. Their mass is approximately 100 times larger than the proton mass.

The important conclusion is that at early epochs the cosmological plasma was dominated by relativistic matter so the equation of state was $p = \rho/3$. It is called radiation dominated plasma (RD-plasma). According to the standard cosmology the domination of relativistic matter lasted for $10^3 - 10^4$ years. After this time the radiation energy density became smaller than the energy density of heavy particles (nucleons and dark matter). It is assumed in classical cosmology that the relativistic gas state equation is valid also for $t < 10^{-10}$s down to $t \approx 10^{-43}$s and up to $T \approx m_{Pl} \approx 10^{19}$GeV. This conclusion is not well established however, as our knowledge of particle physics is experimentally based only up to the energies about 100 GeV. Theoretically, particle interactions are asymptotically free i.e. they weaken as the energy grows. Even if interactions do not disappear at high energies the expansion law should not be drastically altered when the plasma is in the state of thermodynamical equilibrium.

1.7 Freezing of relic particles concentration

As it was stated above, the primordial plasma is typically in thermal equilibrium. This means that all particles with $m \leq T$ are present in the plasma with equal number density per each spin

degree of freedom (forgetting about the coefficient 3/4 differrenti-
ating fermions and bosons as far as particle number is concerned,
and 7/8 for energy density). What happens to those particles during the expansion and cooling down? As the density decreases the
reaction rate falls and at a certain moment the equilibrium is broken. There are two possibilities at this moment. The first is that
the particle interactions are weak and they effectively switch off at
$T > m$. Neutrinos present a good example of that, they decouple
from the plasma at $T \approx 3 - 5$ MeV, while their mass is smaller than
100 eV. Their concentration today is close to that of relic photons
$N_\nu \approx N_{\bar{\nu}} \approx (3/22)N_\gamma$. The value of N_γ is known from observations
$N_\gamma \approx 550(T/3K)^3$, where $T = 2.7 - 3K$ is the background radia-
tion temperature. Hence the relic neutrino concentration is about
$75(T/3K)^3 \mathrm{cm}^{-3}$ for each neutrino type.

The mysterious coefficient 3/22 will now be explained. One
expects that $N_\nu/N_\gamma = 3/8$ in thermodynamical equilibrium (the
neutrino is assumed to have one spin state, see sec.5 ch.4). For $T <$
3 MeV, however, there is no equilibrium with respect to neutrinos
because the reactions with neutrinos are slow. When T drops below
$m_e \approx 0.5$ MeV electron positron pairs annihilate. This leads to
a larger photon density N_γ, while N_ν remains unchanged. The
annihilation proceeds adiabatically so entropy is conserved. The
number of particle degrees of freedom in the primeval plasma (Q in
eq.(1.5)) is $2 + 4 \times (7/8) = 11/2$ before the annihilation and is 2
afterwards. Accordingly, the photon number density increases by a
factor of 11/4. This leads to the ratio $N_\nu/N_\gamma = 3/22$.

Strongly interacting particles behave differently. They remain
in equilibrium with the plasma at $T < m$ and their concentration
follows the equilibrium law $N \approx (mT)^{3/2} \exp(-m/T)$. If the formula were valid today there would be no trace of early hot stages
left. However, the concentration decrease does not continue forever.
At a certain moment the concentration becomes so low that even
strong interactions cannot maintain the equilibrium. After that the
only factor changing the concentration is the expansion of the Universe and the number of particles in a comoving volume remains
constant. This phenomenon is called concentration freezing. As a
result, stable relics of Big Bang can survive until now. The con-

centration of unstable particles (i.e. particles with lifetimes smaller than the age of the Universe, $\tau < 10^{10}$ years) ultimately tends to equilibrium because their production and decay rate is constant, while the expansion of the Universe slows down.

As it is known the equilibrium concentration of particles in an ideal gas is determined not only by their mass and temperature but also by their chemical potential. The latter must be introduced in order to take into account charge (or charges) conservation in particle reactions. For example in an electrically neutral plasma consisting of electrons, positrons and photons the particle energy distribution function is given by:

$$n_{e^+} = n_{e^-} = n_\gamma = \exp(-E/T)$$

(quantum Fermi and Bose effects being neglected).

Obviously this formula is not valid if the total charge of the plasma is non zero i.e. $n_{e^+} \neq n_{e^-}$. Electric charge is conserved so no processes can equalize e^+ and e^- number concentration. This fact is taken into account in equilibrium distribution functions by introducing the multiplier:

$$n_{e^-} = \exp\{(\mu_{e^-} - E)/T\}$$

μ is called the chemical potential of electrons. Let us note that photon number is not conserved so $\mu_\gamma = 0$.

There are baryons (protons, neutrons) in the Universe but there are no antibaryons according to observations. This means that the baryon number (baryonic charge) density is not zero so the baryon equilibrium distribution is described by a non-zero chemical potential μ_b [2]. Neutrino chemical potential is probably also non-zero and of the same order as μ_b. But in contrast to baryons the neutrino chemical potential is not important since neutrino number density is high at $T < m_\nu$.

Chemical potentials appear to be additional parameters on which relic particle concentration depends. Theoretically, they can

---

[2]Particle physics predicts baryon non-conservation at high temperatures. Hence the notion of chemical potential is sensible when baryonic charge is approximately conserved. This is definitely true for $T < 1$ GeV.

be calculated (see ch. 8 on baryosynthesis) but the particle interaction properties in the essential region of energies are not well established so no exact numerical value can be given today.

If it is assumed that the chemical potential of some stable particles (let us call them a) is zero then their contemporary concentration (or frozen concentration) can be easily calculated. It is determined by the condition of thermodynamical equilibrium. The final result for $\mu = 0$ may be considered as a lower bound for the concentration with arbitrary chemical potential.

The latter is determined by the following approximate formula:

$$N_a/N_\gamma = (v\sigma_{aa}m_a m_{Pl})^{-1} \qquad (1.6)$$

where $N_\gamma = 550(T/3K)^3 cm^{-3}$ is the relic photon number density, m_{Pl} is the Planck mass, σ_{aa} is the cross section of aa annihilation, and v is the relative velocity of a and $\bar{a}$ in annihilation.

These considerations have been used by Ya.B.Zeldovich, L.B.Okun and S.B.Pikelner (1965) to estimate the relic quark [3] concentration if quarks could exist as free particles. According to their estimates the abundance of free quarks in nature should be as high as that of gold in disastrous contradiction with quark searches. This result does not depend on quark masses. It is a weighty argument in favour of quark confinement. Quarks are completely new objects: they exist inside elementary particles but cannot be pulled out (see ch.4).

Using formula (1.6) for baryons one obtains:

$$N_b/N_\gamma \approx 10^{-19}$$

which gives the amount of baryons 9 to 10 orders of magnitude less than observed [4]. This once more shows that μ_b is not equal to zero.

Using essentially the same considerations Ya.B.Zeldovich and M.J.Khlopov (1978) calculated the abundance of relic magnetic monopoles (particles carrying elementary magnetic charge). The

---

[3] What quarks are will be explained in chapter 4.

[4] The antibaryon concentration in cosmic rays is about 10^{-4} of that of baryons. They are, however, of secondary origin. The relic antibaryon concentration in the case of baryonic excess is exponentially suppressed.

theory predicts that monopoles should appear in the course of a phase transition during the cooling down of the Universe. Their frozen concentration calculated in this way is unacceptably high. The inflationary model solves this difficulty since if the exponential expansion took place after the monopoles appeared then monopole concentration is exponentially small.

If the concentration of relic particles in the Universe today is known, one can find restrictions on their properties and in particular on their mass. The energy density of such particles should not be too high because otherwise the age of the Universe would be too small. The relation between the matter density and the age of the Universe is:

$$t_U \approx \frac{1}{H}\left(1 + \frac{1}{2}\sqrt{\Omega}\right)^{-1} \tag{1.7}$$

where $\Omega = \rho_{tot}/\rho_c > \rho_a/\rho_c$. An upper bound on the mass of any type of light neutrino was thus obtained $m_\nu < 30$eV (S.S.Gershtein and Ya.B.Zeldovich). Cosmology also puts a lower bound on heavy neutrino mass (if it exists) $m_L > 4$ GeV [3, 49, 54, 64]. This is connected with the fact that the $L\overline{L}$ annihilation cross section grows with m_L like m_L^2 and according to formula (1.6), its energy density decreases.

Arguments of this kind are now widely used to extract information about new particles which are predicted by modern theories but are beyond the scope of the present experimental facilities.

1.8 Nucleosynthesis

The period from $1s$ to $200s$ from the beginning plays an important role in the history of the Universe. At this time light nuclei have been produced: He^4 (25 percent), H^2 (3×10^{-5}), He^3 (2×10^{-5}), Li^7 (10^{-9}) i.e. the usual matter started to appear. Heavier nuclei, however, have been produced much later in stars. There is not enough time for their production at the early stage.

Nuclei output due to primordial nucleosynthesis is sensitive to the value of (n/p) ratio in this period. When $t < 1s$ and correspondingly $T > 1$ MeV it is described by the equilibrium formula:

$$N_n/N_p = \exp(-\Delta m/T) \tag{1.8}$$

where $\Delta m = 1.3$ MeV is the $n - p$ mass difference. The equilibrium is maintained by the weak interaction processes: $n + \nu \leftrightarrow p + e^-$, $n + e^+ \leftrightarrow p + \bar{\nu}$ etc.

When the temperature fell to $T_f = 0.7$Mev these processes became too slow and the (n/p) ratio froze:

$$(N_n/N_p)_f = \exp(-\Delta m/T_f) = const \tag{1.9}$$

At this stage protons and neutrons existed as free particles (not bound in nuclei). Later when T became smaller than 100 keV most of the neutrons (except for those that decayed) became bound in deuterium due to the process:

$$p + n \rightarrow H^2 + \gamma.$$

Deuterium in turn capturing nucleons in the plasma produced He^3 and H^3, which later on with one more proton or neutron formed He^4. At this point practically all existing neutrons were bound. As there are no stable nuclei with $A = 5$ and the plasma was not dense enough, processes like $He^3 + He^4 \rightarrow Be^7$ or $3He^4 \rightarrow C^{12}$ and so on were very rare.

The relative mass abundances of the produced He^4 and H^2, He^3, Li etc. as functions of baryon density in primeval plasma, are shown in figure 1.1. The decrease of deuterium abundance with increasing ρ_b can be explained by a larger probability of deuterium collisions with baryons at higher densities. Only a very small amount of deuterium avoids being transformed to He^4. Furthermore, a higher baryon density results in less deuterium surviving. Deuterium number density in the Universe appears to be a very sensitive indicator of the total amount of baryons. However during the evolution of the Universe the amount of deuterium has probably changed. It is known that the primordial deuterium burns in stars forming He^3. Therefore the combined data on H^2 and He^3 (tritium is radioactive and decays quickly) is needed in order to find bounds on ρ_b.

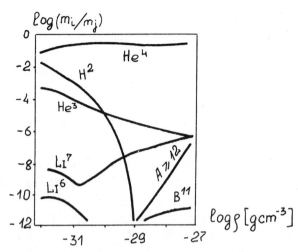

Figure 1.1: Mass fractions of different light nuclei produced during primordial nucleosynthesis as functions of baryon-photon ratio N_B/N_γ.

These data show that the mass fraction of primordial deuterium and helium-3 is about 10^{-4} relative to the total baryonic matter. Hence it follows that the baryonic energy density is bounded from above $\Omega_b < 0.1$.

The agreement of the nucleosynthesis theory with observations (by the abundances of H^2, He^3, He^4, Li^7) is one of the foundations of the hot Universe model. Moreover this agreement allows us, to draw nontrivial conclusions in elementary particle physics. In particular, a restriction on the number of different neutrino species can be obtained. This idea comes from V.F.Schwartzman (1968). Three types of neutrino are known: electronic, muonic and taonic. Cosmology allows not more than an extra two (or maybe three) neutrino species. The arguments are as follows. As it has already been said the He^4 concentration is determined by the N_n/N_p ratio (1.9). The freezing temperature T_f depends on the ratio between the expansion speed $H = 1/2t$ and the rate of the weak interaction reactions, like $\nu + n \leftrightarrow p + e^-$, $e^+ + n \leftrightarrow p + \bar{\nu}$. The latter is determined by the product of, say, the neutrino number density N_ν and the reaction cross section σ. At small energies the cross section is proportional to the energy squared and N_ν is proportional to

T^3. Hence in equilibrium the thermally averaged value $\langle \sigma N_\nu \rangle$ is proportional to the fifth power of temperature.

According to formulae (1.4) and (1.5) the rate of temperature decrease with time depends on number of particle species Q in the following way:

$$T = \left(\frac{45}{16\pi^3} \frac{1}{QGt^2} \right)^{1/4} \tag{1.10}$$

Each type of neutrino adds $7/8$ to Q. Clearly, the larger the number of neutrino types k_ν, the faster the plasma cools down, and the bigger T_f is. Accordingly, as k_ν grows the frozen (n/p) ratio becomes bigger, and more of primary He^4 is produced. During the evolution of the Universe He^4 is not destroyed but can only be produced so the data on the present day abundance of He^4 permits us to derive an upper bound on the amount of primordial He^4. Unfortunately the He^4 concentration in the Universe is not well known. The data varies from 22-23 percent in mass to about 30 percent. The regions with large He^4 concentration are contaminated with heavier elements. So they are probably of secondary origin due to late nucleosynthesis in stars. On the other hand the estimates of the invisible He^4 concentration are not reliable so the quoted value of (22-23 percent) is hardly the proper upper bound. It seems that the safe upper bound on the mass fraction of the primordial He^4, accepted in the literature, is 25 percent.

The calculated abundance of He^4 depends, though rather slightly, on the nucleon number density (see fig. 1.1) or to be more exact, on the ratio N_b/N_γ. This dependance originates from the fact that helium production proceeds in two-body nucleonic collisions. Observations do not give an accurate value of the N_b/N_γ ratio. One can deduce an upper and a lower bound on it by using the data on deuterium and He^3 abundances, because the latter are very sensitive to the baryon number density.

The conclusion that $k_\nu < 4$ can be thus obtained. More cautious authors think however that the data on light nuclei abundances are more uncertain and these abundances might have changed since nucleosynthesis. They consider $k_\nu < 6$ as a safe upper bound.

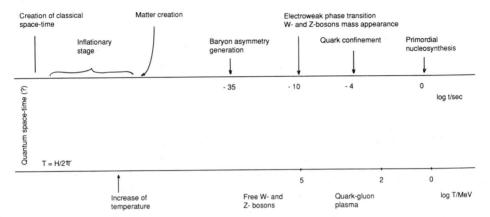

Figure 1.2: The main epochs in the evolution of the early Universe.

Restrictions on other particle types could also be obtained in this way if the particles were present in the plasma during nucleosynthesis ($t = 1 - 100s$, $T = 1\text{MeV} - 100\text{keV}$). In particular, the existence of a mirror or shadow world, fully symmetric to ours, except for gravitational interaction, can be ruled out. The existence of such a world before inflation, however, is not forbidden. It is possible that the mirror particle concentration was strongly diluted due to inflation and their influence to the nucleosynthesis can be neglected. In particular, the superstring theory predicts the existence of such a world at early stages of the evolution of the Universe.

An upper limit on the number density of long lived particles decaying into $p\bar{p}$ pairs during nucleosynthesis or later is another example of applying cosmology to particle physics. Antiprotons born by such decays would interact with He^4 producing deuterium in anomalously large amounts. This is how the strongest limits on new particles, predicted by supersymmetry, were obtained.

The contents of this chapter is illustrated on figure 1.2.

Chapter 2

Problems of classical cosmology

The cosmological model described in the previous chapter has withstood the observational tests. It is confirmed by the existence of background radiation, light nuclei abundances and at last but not least by the theory i.e. by general relativity. In this chapter we will discuss some conceptual difficulties of this model and consider possible ways to resolve them. Let us stress that the modifications needed to solve these problems must be consistent with the standard model which is not to be destroyed by the modern theoretical development.

2.1 Expansion law and equation of state

General relativity equations permit us to find explicitly the expansion law of the Universe if energy density ρ, and pressure p, are known (in a homogeneous and isotropic model). The energy density can be defined by the parameter Ω and the pressure is given by the equation of state $p = p(\rho)$. Let us note that the notion of the equation of state is valid only in relaxed systems when pressure is explicitly determined by energy density. Examples when this is not true will be given later.

The expansion of the Universe is described by scale factor $a(t)$, which characterizes distance between objects as a function of time. The spacetime interval can be written as:

$$ds^2 = dt^2 - a^2(t)[dr^2 + f(r)(d\theta^2 + \sin^2\theta d\phi^2)] \qquad (2.1)$$

where the function $f(r)$ depends on topological properties of the Universe as a whole. $f(r) = r^2$ for a spatially flat Universe (with the Euclidean geometry on average), $f(r) = \sin^2 r$ for a closed Universe and $f(r) = \sinh^2 r$ for the open one. 'r' is a dimensionless parameter while $a(r)$ has the dimension of length.

The $a(t)$ dependence is described by the Einstein equations which in the isotropic and homogeneous case have the following form:

$$\frac{1}{2}\left(\frac{da}{dt}\right)^2 - \frac{4\pi G\rho a^2}{3} = -k/2 \qquad (2.2)$$

$$\frac{d^2a}{dt^2} = -\frac{4\pi G}{3}(\rho + 3p)a \qquad (2.3)$$

where k equals 1, -1 and 0 for respectively closed, open and flat Universe.

If, moreover, the dependence $p = p(\rho)$ is known then all three unknown functions a, ρ and p can be found.

Let us note that equations (2.2) and (2.3) lead to the following law for the energy density variation in the expanding Universe:

$$\frac{d\rho}{dt} = -3H(\rho + p) \qquad (2.4)$$

where $H = \dot{a}/a$ is the Hubble constant. The change in energy density is caused by two factors; first by the expansion of the Universe and second by the work of the pressure forces.

Let us discuss different expansion laws corresponding to different forms of equations of state for the particular case of $\Omega = 1$ ($k = 0$ in eq.(2.2)). As it was already mentioned for an ideal relativistic gas $p = \rho/3$. In this case the scale factor depends on t as:

$$a(t) = a_0(t/t_0)^{1/2} \qquad (2.5)$$

In the Friedmann cosmology this expansion law is assumed to be valid from the "beginning" to about $10^{11}s$ when nonrelativistic particles began to dominate the energy density. As one can easily see the energy density of a relativistic gas decreases as a^{-4} (a^{-3} for the particle number density and a^{-1} for the redshift). Nonrelativistic matter energy density scales as a^{-3} so it must begin to dominate at a certain moment. It is easy to show that for $\Omega = 1$ and $H = 50$ km/s/Mps it takes place at a z-factor approximately equal to:

$$z_m = a_0/a(t) - 1 = \rho_c/\rho_\gamma \approx 10^4$$

where $\rho_\gamma = 4.5 \times 10^{-34} \text{g/cm}^3$ is the relic radiation energy density [1], and ρ_c is the critical energy density, $\rho_c = 5 \times 10^{-30} \text{g/cm}^3$. The photon gas temperature at the transition time was $T = z \times 3\text{K} \approx 30000\text{K} \approx 3$ eV. According to formula (1.3) it corresponds to $t = 10^{11}s$.

After that moment the expansion is determined by the nonrelativistic gas state equation with $p \ll \rho$. With good accuracy one can put $p = 0$. In this case:

$$a(t) = a_1(t/t_1)^{2/3} \tag{2.6}$$

The exponential expansion corresponds to the $p = -\rho$ case:

$$a(t) = a_0 \exp\left(\sqrt{\frac{8\pi G\rho}{3}}\, t\right) \tag{2.7}$$

According to equation (2.4) the energy density is constant i.e. energy decrease due to expansion is compensated by the work of pressure forces.

The main difficulties of the classical cosmology are connected with the very slow growth of the scale factor ($a \approx t^{1/2}$ or $t^{2/3}$). In other words the scale factor $a(t)$ or the size of the Universe is too large at small t. This will be discussed in detail later.

All problems mentioned above can be successfully solved by the inflationary model. This model does not alter the classical cosmology. It only fixes the initial conditions. After the exponential expansion which ended close to the Planck time the expansion law

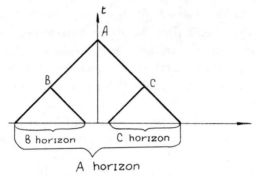

Figure 2.1: Events horizons in the Friedman Universe. Events B and C are not causally connected.

becomes a power law and we return to the standard scenario described for example in the books by Zeldovich and Novikov (1975) or Weinberg (1975).

2.2   The horizon problem

As it was already mentioned the relic radiation coming from different parts of the sky is exactly the same. It practically does not interact with cosmic matter today. The Universe became transparent to the relic radiation after hydrogen recombination, when protons and electrons bound into neutral hydrogen atoms which almost do not interact with long wave photons. The hydrogen recombination temperature is 3,000K which corresponds to $z_r = 10^3$ or $t_r = 10^{12} - 10^{13}$ s. Since that time relic radiation almost did not interact with anything. The size of the causally connected region at the moment of recombination (horizon size) is about $c \times t_r$ (more accurate formula will be given below). Different parts of the sky with the angular size $\theta = (1 + z_r)(t_r/t_0) = 10^{-2}$ should not "be aware" of each other. Nevertheless the relic radiation is identical everywhere. This mystery for the Friedman cosmology phenomenon is called the horizon problem (fig. 2.1).

The intergalactic hydrogen is reionized now by cosmic rays but this reionization took place rather late and in the standard model has had little influence on the relic radiation. As a matter of fact an increase of $\delta T/T$ rather than radiation smoothing is expected because of that. Early reionization models are discussed in connection with the problem of the Universe structure. In this case initially small relic radiation fluctuations can be somewhat smoothed out.

The exact formula for the horizon size takes into account the expansion of the Universe. Let us consider a light ray coming from a point in space to the observer. Light rays propagate along a geodesic for which one can put $d\phi = d\theta = 0$ in eq. (2.1). Then the equation describing the ray propagation is:

$$dr = -dt/a(t)$$

and the distance between the source and the observer is:

$$l = a(t) \int_0^t dt'/a(t') \tag{2.8}$$

Thus $l = 2t = H^{-1}$ for expansion law (2.3) (relativistic matter) and $l = 3t = 2H^{-1}$ for the expansion of nonrelativistic matter (2.6).

Let us note that $l \sim t$ and grows faster than scale factor $a(t) = t^{2/3}$ (or $t^{1/2}$). Any particle will be inside the horizon in the future. It is just the opposite in the case of exponential expansion when $a \sim \exp(H_0 t)$: no signal can ever be exchanged between particles at a distance longer than H_0^{-1}. But if a region of size l_0 was for some reason causally connected its size becomes exponentially large in time, $l(t) = l_0 \exp(H_0 t)$. Although different parts of this region cease to interact as soon as distance between them reaches H_0^{-1}, they "remember" their past. The solution of the horizon problem is based on this fact. (See fig. 2.2).

2.3 The flatness problem

This problem is based on the very close proximity of ρ to ρ_c at early stages of the evolution of the Universe. If at the Planck Time there was $(\Omega - 1) = O(1)$ the Universe would either collapse within $10^{-43} s$ if it is closed or it would expand so fast that no stars could form (if

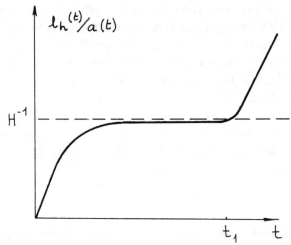

Figure 2.2: The ratio between horizon size and scale factor during the evolution of the Universe.

it is open). In the latter case matter density in 10^{10} years would be much smaller than that observed ($\rho \approx 10^{-29}g/cm^3$).

Let us estimate the necessary fine tuning of Ω. Using the definitions $H = \dot{a}/a$ and $\rho_c = 3H^2/8\pi G$ we can rewrite the equation (2.2) as follows:

$$H^2(1 - \Omega) = -k/a^2 \qquad (2.9)$$

With this equation one can now express Ω as a function of redshift z and its present day value Ω_0:

$$\frac{1 - \Omega}{\Omega} = \frac{1 - \Omega_0}{\Omega_0} \frac{1}{(1 + z)^n}$$

where $n = 1$ for the matter dominated Universe and $n = 2$ for the radiation dominated Universe. Let us estimate how close to unity Ω should have been at early stages, assuming that $(1 - \Omega_0)/\Omega_0 \approx 1$. The results are shown in the table and demonstrate the necessity of surprisingly accurate fine tuning. Note that the earlier time, when the initial conditions are fixed, the closer Ω must be to unity in order to obtain the presently observed Universe. These considerations can be illustrated by comparing the horizon size at the time of the birth

table 1

age of the Universe	period	$\|1 - \Omega\|$
$2 \cdot 10^{10}$ years	contemporary	≤ 1
10^5 years	recombination	10^{-3}
1 sec.	beginning of nucleosynthesis	10^{-16}
10^{-5} sec.	quark-gluon plasma	10^{-21}
10^{-10} sec.	electroweak phase transition	10^{-26}
$t_{Pl} \approx 10^{-43}$ sec.	Universe creation	10^{-60}

of the Universe with the curvature radius in the Friedmann model. The curvature radius R of the homogeneous and isotropic Universe is defined by the trace of the three dimensional curvature tensor:

$$C_G = k/a^2 = 1/R^2$$

and is equal to the scale factor $R = a$. Since the horizon size is equal to $h = ct$ and $a \approx (t/t_0)^{1/2}$ so $h \ll R$ for $t \ll t_0$. It means that the Universe was almost exactly flat when it was born.

We will present some additional simple calculations to illustrate the exactness of fine-tuning of the initial conditions demanded by the classical cosmology. We assume that the Universe was born at $t = t_{Pl} = 10^{-43}s$ instead of $t = 0$. Energy density at that time was about m_{Pl}^4. No contemporary theory describes physical processes under such conditions. Quantum gravitational effects must be taken into account at that time but no one now is able to do that. It is thus impossible to draw any reliable conclusions about what happened at $t < t_{Pl}$ but we hope to understand at least qualitatively the evolution, when $t > t_{Pl}$.

It is sufficient for our objective to take t_{Pl} as time of the birth of the Universe. The present curvature radius is assumed to be the the same as horizon size $R \approx 10^{28}$cm (strictly speaking $R > 10^{28}$cm).

Assuming for simplicity that the expansion law was $a(t) \approx \sqrt{t}$ until the present time we get for the curvature radius at the Planck time:

$$R \approx 10^{28} \text{cm} \sqrt{(10^{-43} s / 10^{17} s)} \approx 10^{-2} \text{cm}$$

and the horizon size at that moment was $l_{Pl} = 10^{-33}$cm. This means that at the Planck time the Universe was flat with 10^{-31} accuracy. Let us do the calculations from the other end. Let the Universe be born at $t = t_{Pl}$ with the naturally expected curvature radius $\approx l_{Pl}$. What would it be like today, at $t_0 = 10^{10}$ years? Calculations show that the contemporary curvature radius R would be $R \approx 10^{-2}$cm.

Can it be that our arguments referring to the curvature radius, size of the Universe and horizon size are wrong ? No, they are right. The same conclusions are drawn when comparing energy densities now and at the Planck time. Density of matter with the equation of state $p = \rho/3$ is known to scale with expansion as a^{-4}. If at the time the Universe was born $a = 10^{-33}$cm and $\rho = \rho_{Pl}$ then the density should have decreased by $(10^{-33}/10^{28})^4$ till today and its value would be $\rho = 10^{-150} g/\text{cm}^3$. The analogous result is obtained when $p = 0$.

The main conclusion is that the classical Friedmann model cannot be extrapolated to very early stages. The law of the scale factor change was different for small t. At that period $a(t)$ must have grown very fast with time. If this is the case, one can answer the questions: why is Ω so close to 1? why is the Universe homogeneous and isotropic? why it is expanding and the expansion law is $v = Hr$?

2.4 First encounter with inflationary model

The essentially new idea which allows to solve all the above-mentioned problems is to introduce negative pressure and to change the equation of state from $p = \rho/3$ to :

$$p = -\rho \tag{2.10}$$

It is sufficient that this law was only approximately valid and only for some finite time. The physical origin of this law will not be discussed yet but its consequences will be considered. As it was already mentioned if $p = -\rho$ then the expansion law becomes exponential (see eq (2.7)). The value of $H = \dot{a}/a$ does not depend on time so eq. (2.10) implies that $|1 - \Omega| \approx \exp(-2Ht)$. Starting from $1 - \Omega = O(1)$ we obtain a flat Universe with the accuracy of about 10^{-60} after 70 Hubble times, $t = 70H^{-1}$. The difference between expansion laws (2.7) and (2.5) or (2.6) is that in the first case $\Omega \to 1$ if $t \to \infty$ while in the second Ω goes away from 1.

The horizon problem is simultaneously solved because the causally connected region which at the moment of the creation of the Universe had the size about $l = m_{Pl}^{-1} = 10^{-33}$cm was inflated up to the necessary size of 10^{-3}cm. This region expands later to reach the present horizon size. The inhomogeneities that existed before inflation are stretched out. The observable part of the Universe thus becomes very homogeneous (with suitable choice of the parameters).

According to equation (2.3) not only energy, but also pressure is the source of gravity. This is connected with relativistic effects and for the case (2.10) leads to a new phenomenon of gravitational repulsion. This repulsion might serve as the initial push that led to the expansion of the Universe if equation (2.10) was valid for energy density and pressure in the microuniverse with size m_{Pl}^{-1} which was born as a result of a quantum jump.

The important question is how and when the exponential expansion stopped and the $p = -\rho$ law changed to $p = \rho/3$. The answer depends on physical origin of the former. We will only note that though the form of this equation is exotic it naturally appears in many cases. These are, for example, phase transitions from symmetric to symmetry-broken phases in primordial plasma, scalar field dynamics with very general initial conditions or theories with higher dimensions, $D > 4$. All this will be discussed further on.

2.5   The problem of cosmological constant

The relation $p = -\rho$ was introduced to cosmology by A.Einstein. Noting that his relativistic gravitation theory when applied to cosmology has no static solutions, he added the extra term $\Lambda g_{\mu\nu}$ to the energy momentum tensor, where $g_{\mu\nu}$ is the metric tensor. Λ is called the cosmological constant. This term just corresponds to the relation $p = -\rho$. It became known later that the Universe is not stationary and for this reason the idea of the cosmological term could be abandoned. Furthermore astronomical observations indicate that its value is very small, $\Lambda < 10^{-56} \mathrm{cm}^{-2}$. Now we understand that the cosmological constant or something equivalent in the early Universe must be introduced in order to obtain the relation $p = -\rho$. This is not, however, the real cosmological constant reflecting the vacuum property , but a methastable configuration of a scalar field which fills the expanding space. It is noteworthy that elementary particle physics predicts that real cosmological constant should be non-zero and roughly 100 (!) orders of magnitude greater than the existing limit. A cancellation mechanism of Λ by that degree is not known at present. This fact constitutes one of the most important problems in cosmology as well as in particle physics. A separate chapter is devoted to this problem.

2.6   Baryon asymmetry of the universe

One more cosmological problem, that was solved by elementary particles physics, besides the horizon and flatness problems, is the baryon asymmetry one. Matter exists in the Universe in the form of baryons (protons and neutrons) and does not in the form of antimatter (antibaryons). Let us note that antimatter in the form of antineutrinos should be about the same as that of neutrinos or of the number of relic photons and large in comparison with the number of baryons (there are about 10,000,000,000 photons to one baryon).

It is not necessary to modify the Friedman model to solve this problem. The expansion of the Universe along with the differences between particle and antiparticle interactions (observed in experiments) and the baryon number non-conservation (predicted by the theory) lead to a solution of this problem. Independently on the initial conditions the primordial plasma arrives at $T > m_B = 1$ GeV to the state with only a small excess of baryons over antibaryons $(N_B - N_{\overline{B}})/(N_B + N_{\overline{B}}) \approx 10^{-9} - 10^{-10}$. After the baryon annihilation at $T < m_B$ this excess has led to a Universe with the 100 percent baryonic asymmetry (A.D.Sakharov 1967). Now this is the only experimental proof of baryon non-conservation. Chapter 8 is devoted to this problem. A short introduction to particle physics must be given before we proceed to consider genuine cosmological problems.

Chapter 3

A short introduction to field theory

The basic concepts of field theory essential for the rest of the book are presented in this chapter. This presentation is sufficient as a first encounter with the subject. The book by Landau and Lifschitz (1975) or the first chapter of Bogoliubov and Shirkov (1976) are advised for deeper studies. We chose the historical approach in our presentation to electrodynamics and gravitation instead of the progression from simple things to more complicated ones i.e. from scalar through vector to tensor field. We consider the scalar field that plays an important role in cosmology as well as in particle physics only in chapter 4. The scalar field will also be discussed later along with inflation. Our choice was based on the fact that electromagnetic and gravitational fields are observed in nature while the scalar field is not.

3.1 Fields in special relativity

Historically the first field theory was probably the Newton's theory of gravitation. According to this theory gravity can be described by a scalar function $\phi(r)$ (scalar field) in three dimensional space. This function is called the scalar potential. This theory gives the correct description of gravitational interactions in nature only when the velocities are small ($v \ll 1$). The principle shortcoming of this theory is the idea of instant interaction.

Apart from scalar there can also be vector, tensor etc. fields in nature. Vector fields are defined by three functions V_i in each point of 3-dimensional space. These functions obey special transformation rules when transition is made from one coordinate system to another. This will not be discussed in detail and we will proceed to special relativity in which fields are classified in 4 dimensional spacetime. This classification is also valid in general relativity.

Before the special theory of relativity was formulated, time and space coordinates were discussed separately. Time was assumed to be absolute - it did not vary in transformations from one coordinate system to another $t = t'$. The spatial coordinates changed in a definite way $x \leftrightarrow x'$. The Galileo relativity principle states that physical laws do not change after transition to a coordinate system moving with constant velocity. The coordinate transformation law for such movement along z axis is: $t = t'$, $x = x'$, $y = y'$, $z = z' + vt$. Field classification in three dimensional space is connected with this transformation law.

The Galileo invariance is however only approximate and valid in the case of small v ($v \ll 1$). For velocities close to velocity of light, physical laws also do not change with transformation from one inertial coordinate system to another. Now, however, the form of corresponding coordinate transformation is changed. In particular, time and space coordinates are mixed: $t' = f_1(t, x)$ and $x' = f_2(t, x)$. Therefore we come to the notion of four dimensional spacetime.

Special relativity transformations, called Lorentz transformations, leave the interval invariant:

$$ds^2 = dt^2 - dx_1^2 - dx_2^2 - dx_3^2 \qquad (3.1)$$

The four-dimensional space-time with the distance defined by this expression is called Minkowski space. The Lorentz transformations can be formally considered as rotations in 4 dimensional spacetime with purely imaginary generators in tx_1, tx_2, tx_3 planes. These rotations correspond to physical transformations to coordinate systems moving along one of the axes x_i. Transformations in $x_i x_j$ planes are ordinary rotations. The Lorentz transformations consist of three independent rotations in $x_i x_j$ planes and three independent movements along x_i axis. Together this makes six parameters. Besides these, there are four more transformations allowed in Minkowski space corresponding to a shift of coordinate system centre, $x_\mu \to x_\mu + a_\mu$.

The transformations to a moving coordinate system change the flow of time, and so they are called time-space transformations. They have very important invariance properties. In particular, the light cone $t^2 - r^2 = 0$ is transformed by Lorentz transformations onto itself. World lines of any particles regardless of how they move remain within the light cone. It reflects the fact that movement faster than light is impossible. No Lorentz transformation can move a point outside the light cone if it was inside before. Lorentz transformations cannot change the direction of time (these are or-tochronic Lorentz transformations, sometimes the general Lorentz group including the time reversal transformation is considered but the latter cannot be realized physically).

Let us return to field classification in special relativity. As it was already mentioned Lorentz transformations mix the time axis with space ones so field classification in four dimensions is different from that in three dimensions. A 3-dimensional scalar may form a component of a 4-dimensional vector or tensor. A 4-dimensional scalar however is always a 3-dimensional one. A scalar is sometimes called a zero rank tensor. Vector V_μ (first rank tensor) ($\mu = 0, 1, 2, 3$) in 4-dimensions contains a 3-dimensional vector V_i ($i = 1, 2, 3$) and scalar V_0. The numerical values of V_0 and V_i depend of course on the choice of reference frame, but a square of 4-vector $V_\mu^2 = V_0^2 - V_i^2$ is invariant. A symmetric tensor of second rank in 4 dimensions breaks up into symmetric tensor T_{ij}, vector T_{0j} and scalar T_{00} in 3 dimensions. Let us note that a symmetric tensor is not irreducible. It contains a scalar part, i.e. the trace T_i^i, and the genuinely tensor part, $T_i^j - \delta_j^i T_k^k / 3$. An antisymmetric tensor in 4 dimensions breaks up into two vectors: $H_k = \epsilon_{kij} F^{ij}$ and $E_k = F_{0k}$ in 3 dimensions.

Tensor characteristics of a field in three dimensional space define its spin S. Spin represents angular momentum of a field quantum i.e. intrinsic angular momentum of the corresponding elementary particle. The spin of a scalar field is of course equal to zero. A vector field consists of two parts: scalar with $S = 0$ and vector with $S = 1$. The latter can have three projections of spin on the z axis $S_z = -1, 0, 1$. Physical particles, as experiments show, have pure spin states, but not a mixture of them. So additional conditions excluding smaller spins, are imposed on fields. Probability of some processes with, say, vector fields in quantum field theory can be negative unless spin zero state is eliminated. Usually to this end the condition $\partial_\nu V^\nu = 0$ is imposed in vector field theory and an analogous one for tensor field. Massless field theories like electromagnetism and gravitation, possess additional symmetries, ensuring that low spin states do not appear. This is gauge invariance in electromagnetism. It allows to add a gradient of a scalar function to the electromagnetic potential simultaneously changing the phases of the wave functions of charged particles. It is general covariance in gravitation that allows any change in coordinate system.

Let us note that not only low spin states but also low projections of high spin states are eliminated. It is convenient to choose z axis in the direction of motion of the particle and project spin on it. It ensures that spin projection does not change with time. A massive vector field has three degrees of freedom $S_z = -1, 0, 1$, while a massless has only two $S_z = -1, 1$. The tensor gravitational field has also only two degrees of freedom $S_z = 2, -2$ while low spin projections $S_z = 1, 0, -1$ are absent. The statement of the absence of low spin states is Lorentz invariant because quanta of these fields move with the speed of light. The $S_z = +S$ and $S_z = -S$ states cannot be interchanged by Lorentz transformations if no space inversion is included.

It is easy to establish the equation of motion governing a free scalar field. General considerations show that it should satisfy a second order equation. The only invariant second order operator is the d'Alembert operator:

$$\partial^2 = \frac{\partial^2}{\partial t^2} - \frac{\partial^2}{\partial x^2} - \frac{\partial^2}{\partial y^2} - \frac{\partial^2}{\partial z^2} \tag{3.2}$$

Thus the scalar field should satisfy the equation:

$$\partial^2 \phi = J \tag{3.3}$$

where J is a source of this field. One more term can be introduced to this equation without violating its Lorentz invariance:

$$(\partial^2 + m^2)\phi = J \tag{3.4}$$

Massive field or massive particles (quanta of this field) are described by this equation. It can be easily shown that

$$\exp(-iEt + ikr)$$

is a solution of eqs. (3.3) and (3.4). The only difference is that $E^2 = k^2$ in the first case while $E^2 = k^2 + m^2$ in the second. This is the relativistic formula connecting mass, energy and momentum of a particle.

Usually field theories are formulated in Lagrangian formalism. A Lorentz invariant function of field and its first derivatives called the Lagrangian (or to be more exact the Lagrangian density) is constructed first. This automatically ensures Lorentz invariance of the theory. The action functional is later determined:

$$S = \int d^4x\, L = \int dt d^3x\, L$$

Equations of motion are defined by demanding that the action be extremal i.e. the functional derivative of S should be zero.

$$\frac{\delta S}{\delta \phi} = \frac{\partial L}{\partial \phi} - \frac{\partial}{\partial x_\mu} \frac{\partial L}{\partial \frac{\partial \phi}{\partial x^\mu}} = 0 \tag{3.5}$$

The scalar field Lagrangian is:

$$L_\phi = \frac{1}{2}\left(\frac{\partial \phi}{\partial x_\mu}\right)^2 - \frac{1}{2}m^2\phi^2 \tag{3.6}$$

A scalar field source J can be introduced by adding $J\phi$ to the Lagrangian density. As the Lagrangian density is a scalar quantity, the current J should also be a scalar. Analogously, the vector field

source, introduced by adding $A^\mu J_\mu$ to the Lagrangian, is a vector
and the tensor field source is a tensor.

Equations of motion of high spin fields may contain some additional terms but eventually they all have form (3.3) or (3.4). So far
only integer spin fields with $S = 0, 1, 2$ have been discussed. They
are called boson fields. There can however exist half-integer spin
fields $S = 1/2, 3/2$. These are fermion fields. The important difference between fermion and boson fields is that two identical fermions
can never be in the same state. Bosons behave in exactly the opposite way: if there are already many bosons in a state the probability
for extra boson production in this state is greatly enhanced. The
principle of a laser is based on this phenomenon, which is called
Bose-condensation. The existence of classical boson fields is from
the quantum point of view Bose-condensation of corresponding particles. It is evident that classical fermion fields do not exist. The
more accurate formulation of fermion-boson difference is that the
wave function of a system of particles should be antisymmetric with
respect to identical fermion interchange (Pauli principle), and symmetric with respect to identical boson interchange. This implies, in
particular, the impossibility for two fermions to exist in the same
state and that a bound state of an even number of fermions is a
boson. The Pauli principle and Bose-condensation cannot be obtained as a result of potential interactions of any kind. Classical
gravitational and electromagnetic fields have long been known and
only relatively recently it was understood that there exist quanta
of these fields. Fermions have been at once discovered as particles.

It should be added that fields not only in 4 dimensional spacetime, but also in larger dimensions e.g. $D = 10$ or $D = 26$ are considered in contemporary physics. In correspondence to the above
presented discussion one field in 10 or 26 dimensions contains many
fields in 4-dimensions. It should be noted that the additional $(D-4)$
dimensions are compactified and cannot be observed at small energies $(E \ll m_{Pl} = 10^{19} \text{GeV})$.

3.2   Vector field - electrodynamics

The classical theory of electromagnetic interactions formulated by J.C.Maxwell, was in fact the first relativistically invariant field theory, although it appeared long before Einstein's special relativity. Two three dimensional vectors, representing electric (E_i) and magnetic (H_i) fields, are the basic objects of this theory. Despite the fact that 4- dimensional transformations were not known at that time, it was clear that fields E_i and H_i were correlated. For example if in one coordinate system $H \neq 0$ and $E = 0$ then in another, moving with speed v, there appears electric field $\mathbf{E} = [\mathbf{v} \times \mathbf{H}]$ and the value of magnetic field changes. The reverse statement is also valid: if uniform fields E and H are perpendicular in one coordinate system, then there exists another coordinate system, moving with constant velocity relative to the original one in which either $E = 0$ if $(E < H)$ or $H = 0$ if $(H < E)$. This connection between E and H indicates that they may form one common four dimensional tensor. It is known from the preceding section that this may be an antisymmetric tensor $F_{\mu\nu}$:

$$F_{\mu\nu} = \begin{bmatrix} 0 & E_x & E_y & E_z \\ -E_x & 0 & -H_z & H_y \\ -E_y & H_z & 0 & -H_x \\ -E_z & -H_y & H_x & 0 \end{bmatrix} \tag{3.7}$$

This guess will be justified if one checks that the transformation of E and H when transition from one inertial coordinate system to another is made looks like four dimensional rotations of $F_{\mu\nu}$. It is very often the case that if the coordinate system is changed then calculations are simplified. As an example, let us discuss movement of an electron in perpendicular electric and magnetic fields. If $H > E$ then there exists a coordinate system in which only magnetic field $H' = H'(H, E)$ is present. It is well known that the electron moves along a spiral in a pure magnetic field. Going back to the first coordinate system one can easily find the electron trajectory. The procedure is the same in the case when $E > H$. The case $|\mathbf{E}| = |\mathbf{H}|$ and $\mathbf{EH} = 0$ corresponds to the fields that are the same in all coordinate systems i.e. to electromagnetic wave.

It is now clear that E and H form a second rank tensor but the electromagnetic field is still called a vector field and the reason is not simply historical. The equations for F or equivalently for E and H are:

$$F_{\mu\nu,\lambda} + F_{\lambda\mu,\nu} + F_{\nu\lambda,\mu} = 0; \quad F^{\mu\nu}_{,\nu} = -4\pi j^{\mu} \qquad (3.8)$$

in four dimensional convention. Now and later on the compact notation $f_{,\mu} = \partial f / \partial x^{\mu}$ is used. These two Maxwell equations in three dimensional convention look as:

$$\text{divH} = 0, \quad \text{curlE} = -\frac{1}{c}\frac{\partial H}{\partial t}, \quad \text{curlH} = \frac{4\pi}{c}\frac{\partial E}{\partial t} + \frac{4\pi}{c}j, \quad \text{divE} = 4\pi\rho$$

where ρ is electric charge density and j is electric current vector.

The six quantities E and H are not independent. Four equations divH $= 0$ and curlE $= -(1/c)\,\partial H/\partial t$ impose four constraints and this leaves only two independent quantities. Let us introduce a four dimensional vector A_{μ} which is connected to the tensor $F_{\mu\nu}$ by the relation:

$$F_{\mu\nu} = A_{\mu,\nu} - A_{\nu,\mu} \qquad (3.9)$$

Thus F is antisymmetric, the first equation (3.8) is automatically satisfied and six components of E and H can be expressed through four components of the vector A_{μ}. That is the reason to call electromagnetic field a vector field. Condition (3.9) does not determine A_{μ} unambigously. A gauge transformation can still be done i.e. the gradient of a scalar function can be added to A:

$$A_{\mu} \rightarrow A_{\mu} + \partial_{\mu}f \qquad (3.10)$$

This transformation does not change $F_{\mu\nu}$ and thus leaves the field equations invariant. One more condition can be imposed on the free field A_{μ} so it has only two independent components.

Electromagnetic field can be bound to charges, like for example the Coulomb field, or it can be free like an electromagnetic wave. These are radio waves, light (high frequency electromagnetic waves) etc. In Maxwell theory there are no constraints on the electromagnetic wave frequency. One of the field equation solutions is $E_x = H_y = u(x - t)$, where u is an arbitrary function. In particular $u = \cos[\omega(x - t)]$ is a solution. As frequency ω can be arbitrarily low, the energy of field quanta can be arbitrarily small as well. Vanishingly low frequencies are possible only if the mass of the field is zero. This is the cause why the Coulomb interaction is long-range, $E \approx 1/r^2$. The electric field flux through any closed surface does not decrease with distance. Let us note that static field decreases faster $(1/r^2)$ than radiation field $(1/r)$. Thus, only electromagnetic waves(radio, x-rays, light) are registered in astronomical observations (observations of distant sources) but not static electric or magnetic fields that may dominate the radiation field in the neighborhood of a star. On the other hand, the electric field of a star is negligible not only because of the factor r^{-2} but mainly due to smallness of the star's electric charge. Close to pulsars the electric field can be really strong (up to 10^{14}V/cm) but this is caused by rapid rotation of the pulsar magnetic field which can reach 10^{13}Gs. However, as pulsars have neither electric nor magnetic net charge their fields fall faster than r^{-2} and far from the pulsar are negligible. Charged particles can be accelerated by such fields up to 10^{20}eV (the characteristic size of pulsar field of order of 10^6cm).

There is one more very important property of the Maxwell equations. The equations themselves impose a condition on the field source. Let us consider the first pair of the Maxwell equations:

$$\mathrm{div}E = 4\pi\rho; \quad \mathrm{curl}H = \frac{4\pi}{c}\frac{\partial E}{\partial t} + \frac{4\pi}{c}j$$

Applying div operator to the second equation and substituting the first one into the result, we obtain the following relation between charge density and electric current:

$$\frac{\partial\rho}{\partial t} + \mathrm{div}\,j = 0 \tag{3.11}$$

Therefore currents and charges cannot be chosen arbitrarily. They should satisfy equation (3.11) which is called the charge conservation law.

The important thing is that the charge conservation is a consequence of the Maxwell equations. Numerous tests of the Maxwell electrodynamics show no deviation from the theory. Hence electric charge must be conserved. In other words the question about charge non-conservation in the framework of the Maxwell theory is senseless.

Conservation law (3.11) can be written in 4-dimensional form if one takes into account that 3-dimensional vector j_k and scalar ρ form 4-dimensional vector j_μ:

$$\partial_\mu j^\mu = 0 \tag{3.12}$$

It is enough to look at the second equation (3.8) to see that j_μ is a four vector. Four-divergence of a second rank tensor is always a four vector.

Equations of electrodynamics are often written in terms of A_μ instead of $F_{\mu\nu}$. They have a very simple form if the condition $\partial_\mu A^\mu = 0$ is imposed thanks to gauge freedom (3.10). In this case one vector equation:

$$\partial^2 A_\mu = 4\pi j_\mu \tag{3.13}$$

is equivalent to all the Maxwell equations. Note that it has the same form as equation (3.3) for scalar field.

The theory is not completely defined by equations (3.8) or by equivalent equations (3.13). Given electric current j_μ, $F_{\mu\nu}$ or A_μ can be calculated. Apart from that one needs to know how to calculate particle motion in an external electromagnetic field. To this end the equations of motion for particles with electromagnetic field interaction taken into account must be formulated. The form of electromagnetic interaction is determined by the principle of gauge invariance. Namely the theory must be invariant with respect to transformations (3.10), accompanied by a phase rotation of the charged field, $\psi \to \exp(ief)\psi$ (e is the charge of the field). Hence it follows that electromagnetic interactions can be introduced by the simple procedure of change of derivatives. All usual derivatives in the free field equations must be replaced by the so called covariant derivatives D_μ:

$$\partial_\mu \to D_\mu = \partial_\mu - ie A_\mu \tag{3.14}$$

This is how the interactions are basically introduced in particle physics. The same idea is used in gravitation. Let us now construct the energy-momentum tensor of the electromagnetic field. This tensor is the source of the gravitational field. The procedure of deriving the energy momentum tensor is described in books mentioned in the beginning of this chapter. The result is:

$$T_{\mu\nu} = \frac{1}{4} g_{\mu\nu} F^{\alpha\beta} F_{\alpha\beta} - F_{\mu\alpha} F^{\alpha}_{\nu}$$

Let us note that $T_{\mu\nu}$ is conserved:

$$\frac{\partial T^{\nu}_{\mu}}{\partial x^{\nu}} = 0$$

One can easily prove this using the Maxwell equations. The energy momentum tensor of electromagnetic field has one more interesting property:

$$T^{\nu}_{\nu} = 0$$

This condition is a consequence of the conformal invariance of electromagnetic field theory. It also implies the equation of state of a photon gas:

$$p = \rho/3$$

There is a very nice phenomenon connected with gauge invariance of electromagnetic interactions. The phase of the wave function of a particle moving in an electromagnetic field changes by:

$$\Delta\phi = e \int A_{\mu} dx^{\mu}$$

The integral is calculated along the path of the particle. This expression is consistent with gauge transformation of a wave function discussed above. Let us consider the interference of particles moving along two different paths in magnetic field $A_0 = 0$, $A_i \neq 0$ (see fig. 3.1).

The interference picture is determined by the phase diffrence:

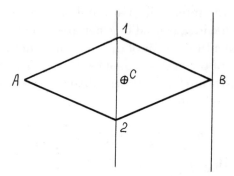

Figure 3.1: Interference experiment demonstrating the Bohm-Aharonov effect. A1B and A2B are two different paths of electrons, C is a solenoid with magnetic field inside.

$$\Delta\phi_{12} = e \int A dx$$

The integral is calculated along closed path A1B2A. Thus an observable effect depends on the seemingly unphysical vector potential A_μ. The value $\Delta\phi_{12}$ is clearly gauge invariant. Transformations (3.10) cannot change it. This can also be proved using the well known theorem of vector analysis:

$$\int A dx = \int \text{curl} A \ ds = \int H ds$$

That is the integral of the vector potential circulation around a closed curve is equal to the magnetic flux through a surface enclosed by that curve. It seems that everything is all right because the effect is expressed in terms of magnetic field H. The experiment could however be arranged in such way that the magnetic field would be confined in a long and thin solenoid far away from particle paths. Nevertheless, the interference pattern should be sensitive to the field inside the solenoid. This phenomenon, called the Bohm-Aharonov effect, has been observed in full agreement with the theory.

Maxwell equations (3.8) or (3.13) are always right from the microscopic point of view but equations of macroscopic electrodynamics can be different. There are polarization effects in the medium that should be taken into account. In principle these equations can be derived from the first principles i.e from equations (3.8) and the fundamental laws describing medium properties. Practically, these properties are characterized by phenomenological dielectric and magnetic constants, etc. Macroscopic electrodynamics deals with light dispersion and refraction, Cherenkov radiation and many other phenomena.

It appears, however, that the vacuum is also polarized by external fields. It is connected with production and annihilation of pairs of charged particles interacting with the field. This simple description is not accurate however. The particle production and annihilation in vacuum does not mean that the vacuum is not stationary. The external field just modifies stationary wave functions. This can be understood better if we consider a hydrogen atom as a simplified example. The atom can be in one of its stationary states described by the wave function:

$$\Psi = \Psi_\alpha e^{-i\omega_\alpha t}$$

Any superposition of such states is not stationary and it radiates electromagnetic waves. But in an external field a stationary wave function of the atom can have the following form:

$$(C_1\Psi_1 + C_2\Psi_2 + ...)\exp(-i\omega' t)$$

It does not mean that the atom "jumps" from Ψ_1 to Ψ_2 and so on. But if the external field is quickly switched off the atom turns out to be in a non stationary state:

$$C_1\Psi_1\exp(-i\omega_1 t) + C_2\Psi_2\exp(-i\omega_2 t) + ...$$

Thus, there appears a superposition of ground and excited states. This excitation results from the non stationarity of the field (quick switching off).

The same mechanism leads to small corrections to the Maxwell equations in vacuum that do not change general properties of these equations like charge conservation or gauge invariance. The equations become nonlinear and there appears, for example, double refraction (refraction coefficient in magnetic field depends on light polarization). The averaging of equations (3.8) over a material medium or vacuum predicts a number of new phenomena. (Equations (3.8) should be understood as operator equations in the case of vacuum averaging.) Of course for the complete description of a physical system the equations of motion of matter in electromagnetic field are necessary besides the Maxwell equations.

The average values of E and H in homogeneous and isotropic Universe should be zero due to the space symmetry. What is less trivial is that total charge (or mean charge density) of closed Universe should be zero. This follows from the Gauss theorem applied to the Universe divided into two parts by any closed surface.

3.3   Tensor field - gravitation theory

The success of Maxwell electrodynamics inspired physicists to build other relativistic theories along the same lines. At the beginning of the century only one more field was known, the gravitational field. The gravitational interaction has been studied earlier than the electromagnetic one. In particular, the Newton law was discovered before the Coulomb law. Despite that there existed no relativistic theory of gravitation at the beginning of century. Gravitation was described by the scalar potential:

$$\Delta\phi_{Newton} = -4\pi G\rho \tag{3.15}$$

where ρ is the mass density. This equation resembles the electrostatic potential equation:

$$\Delta\phi_{electric} = 4\pi\rho_e$$

The only difference is that the mass density is replaced by the charge density.

There is however a very important difference between electric and gravitational fields, which is not reflected by these equations. This difference is that two electric charges of the same sign are repelled while two gravitational charges of the same sign i.e. two masses are attracted. Thus, the gravitational and electromagnetic field theories should be different. The requirement of positive definiteness of energy in relativistic field theory implies that only even spin (scalar, second rank tensor) theories lead to attraction of the same sign charges, while the odd spin (vector) theories lead to repulsion. Later we shall see that the gravitational field in relativistic theory must be described by either a scalar or a second rank tensor. The relativistic generalization of equation (3.15) consists of replacing the Laplace operator by the d'Alembert operator and of determining the tensor properties of the field source and therefore of the field itself.

The properties of the source $4\pi G\rho$ in eq. (3.15) with respect to Lorentz transformations permit to conclude that there are only two possibilities. Energy density may be the nonrelativistic limit of either the energy-momentum tensor $T_{\mu\nu}$ or scalar T_μ^μ. The gravitational field would be a tensor in the first case and a scalar in the second. It is proved experimentally that the first case is true.

Let us find the energy density transformation after a transition to a moving coordinate system and compare the result with the transformation of four dimensional vector current (ρ, j).

Volume of a moving body decreases in accordance with the expression:

$$V = V_0\sqrt{1 - v^2}$$

where V_0 is the volume of the body at rest. Charge is not transformed and the energy of the moving body increases as:

$$E = m\frac{1}{\sqrt{1 - v^2}}$$

where m is the energy of the body at rest i.e. its mass. Hence the energy density transforms proportionally to the second power of $\gamma = (1 - v^2)^{-1/2}$, i.e.:

$$\rho_m = \rho_{m_0}\gamma^2$$

while the electric charge density variation is proportional to the first power of γ:

$$\rho_e = \rho_{e_0}\gamma$$

There is one more power of γ in energy density transformation law as compared to that of a vector. It means that ρ behaves like 00-component of a second rank tensor.

We know such a tensor already. It is the four dimensional energy-momentum tensor. This symmetrical second rank tensor is usually denoted as $T_{\mu\nu}$. Component T_{00} is energy density and components T_{i0} are energy flux or momentum density. The spatial-spatial components of it form the three dimensional stress tensor. As an example we write down the energy-momentum tensor of a non-viscous fluid:

$$T_{00} = \rho$$

$$T_{0i} = j_i = (\rho + p)v_i$$

$$T_i^j = (\rho + p)v_i v^j - p\delta_i^j$$

For the fluid at rest $(v = 0)$ ρ is the energy density and p is the pressure.

The tensor $T_{\mu\nu}$ should be substituted into the d'Alembert equation (3.3) in order to obtain a relativistic theory of gravity. Correspondingly the gravitational field should be described by a second rank tensor in order to preserve relativistic covariance:

$$\partial^2 \ \Psi_\nu^\mu = 4\pi G T_\nu^\mu \qquad (3.16)$$

Functions Ψ_ν^μ should be known in order to completely determine the gravitational field. Not all ten functions Ψ_ν^μ are independent, however. The conservation law of T_ν^μ which is written as:

$$\partial_\mu T_\nu^\mu = 0 \qquad (3.17)$$

imposes the same conditions on Ψ_μ^ν:

$$\partial_\mu \Psi^\mu_\nu = 0$$

Gravitational field is usually expressed by functions h^μ_ν instead of Ψ^μ_ν:

$$h^\mu_\nu = \Psi^\mu_\nu - \frac{1}{2}\delta^\mu_\nu \Psi^\alpha_\alpha \qquad (3.18)$$

where $\delta^\mu_\nu = diag(1,1,1,1)$ is the unit tensor. Tensor $T_{\nu\mu} = g_{\mu\alpha}T^\alpha_\mu$ is symmetric and therefore $\Psi_{\mu\nu}$ must be also symmetric. The Newtonian gravitational potential $\phi = 2h^0_0$ satisfies the equation:

$$\partial^2 \phi = 4\pi G T^0_0 \qquad (3.19)$$

Another possible relativistic gravitation theory can be formulated if the trace of energy momentum tensor, T^μ_μ, is chosen as the field source. In this case the field should also be scalar. This scalar theory of gravitation is rejected by experiments because electromagnetic waves are deflected in gravitational field despite the fact that $T^\mu_\mu = 0$ for them.

Note that it is very difficult to distinguish the scalar theory of gravity on the basis of stationary non relativistic systems. The difference between the scalar and tensor theories lies in very small and hardly observable effects. The two theories are very similar due to the equation:

$$\int T^\alpha_\alpha dV = m \qquad (3.20)$$

which is valid for stationary systems. To illustrate how it works for the electromagnetic energy-momentum tensor let us imagine a spherical balloon filled with photon gas. The equation $T^\mu_\mu = 0$ holds inside the balloon but equation (3.20) is true because of stresses in the balloon walls which arise because of the gas pressure.

Scalar theory of gravity being formally self-consistent, experiments unambigously confirm the tensor theory. It is verified with very good accuracy of about 0.1 percent in the deviations from the Newton theory, which by themselves are small.

Now the tensor theory of gravity has become a theoretical background in space navigation.

Not to leave any stone unturned, it has to be noted that equations (3.16) are by no means the exact equations of gravity. The reason is very simple. The energy-momentum tensor which is the gravitational field source, is conserved. The gravitational field itself, however, has its own energy which can be transferred from the energy of matter and back again. Thus the energy-momentum tensor of matter is not conserved. One has to include in it extra terms accounting for the energy of the gravitational field. Consequently eqs (3.16) turn into nonlinear equations describing self interaction of gravitational field and modifications of T_ν^μ due to gravity. Equations (3.16) are only a weak field limit of the exact theory.

There is a deep analogy between the theory of gravity and that of vector fields, if the latter are charged and because of that are selfinteracting and obey nonlinear equations. These theories are called nonabelian gauge theories (contrary to abelian electrodynamics) The progress in particle physics in recent years is to a large extent based on these theories (see ch.4).

The interaction between gravitational field and matter is determined by the equivalence principle, which says that all bodies are identically accelerated in a gravitational field. The principle can be reformulated as the requirement that at any given point any gravitational field can be cancelled out by an appropriate choice of the coordinate system. Einstein used this formulation as the starting point of his approach to General Relativity. The simplified theory described above, which is valid for weak gravitational fields is based on the notion of pseudoeuclidean spacetime (Minkowski space) in which gravity is considered as an external field. In principle this can be generalized to the case of strong fields. Flat spacetime is, however, nonobservable and in this sense fictious. All particles and fields universally interact with gravity. There are no physical objects which are neutral with respect to gravity. Hence no reference frame realized with physical bodies describes the geomentry of Minkowski space.

In this formalism however the equivalence principle is hidden, and the speed of light depends on the gravitational field, frequency of light emitted by atoms varies and so does their size. Nevertheless, a consistent gravitational theory in flat spacetime can be formulated and might even be useful in supergravity theory. Such an alternative to Einstein's theory has been developed by A.A.Logunov and coauthors. Ya.B.Zeldovich and L.P.Grishchuk discuss the difficulties of this theory in detail and prove the necessity and consistency of the geometric theory of gravity in their papers of 1986 and 1988. L.D.Faddeev (1982) proved that there are no contradictions concerning energy and momentum of the gravitational field in General Relativity.

Let us return to General Relativity. We are not going to discuss the details of the geometric theory, but nevertheless we would like to write down the basic equations just because of their beauty:

$$R_{\mu\nu} - \frac{1}{2}g_{\mu\nu}R = 8\pi G T_{\mu\nu} \qquad (3.21)$$

where $G = m_{Pl}^{-2}$ is the gravitational coupling constant, $T_{\mu\nu}$ is the energy momentum tensor which serves as the source of gravitational field. On the left hand side of the equation there are only geometric quantities : $g_{\mu\nu}$ is the metric tensor determining the interval in the Riemann space:

$$ds^2 = g_{\mu\nu}dx^\mu dx^\nu$$

and $R_{\mu\nu}$ is the Ricci tensor obtained in a definite way from $g_{\mu\nu}$ and is a function of $g_{\mu\nu}$ and its first and second derivatives. $R = g^{\mu\nu}R_{\mu\nu}$ is called the curvature scalar.

The quantity $h_{\mu\nu}$ introduced above describes the deviation of the metric from the flat space one due to interaction with matter: $h_{\mu\nu} = g_{\mu\nu} - g_{\mu\nu}^0$. Equation (3.16) can be obtained from eq (3.21) by expanding it in powers of h only up to the first order.

The gravitational analogue of gauge invariance in electrodynamics is the principle of general covariance. It reflects the invariance of the theory with respect to the choice of arbitrary coordinate system. This principle forms the basis of General Relativity. In fact, the whole theory can be derived from it.

Let us recall general properties of the tensor theory of the gravitational field which is absolutely equivalent to General Relativity. The tensor theory implies universal attraction. It automatically leads to a covariant conservation law of the energy-momentum tensor (because the usual derivatives are replaced by the covariant ones). If only one kind of particles or only one field (like the electromagnetic) would be the source of the gravitational field, then the energy-momentum conservation law would imply the equation of motion of this particle or field. It is noteworthy that if gravity is neglected one can construct a conserved second rank tensor which conservation law does not imply any conditions on the field. General Relativity is a self-consistent theory, the sources give rise to the field and the field governs their motion. This was the starting point for the idea to reduce all dynamics to geometry and to find a unified geometrical field theory. In such a simple form no attempt has been succesful. One equation of energy momentum conservation is not enough to obtain equations of motion of many different particles and fields. Let us explain this in more detail. For example, during nuclear reactions in uranium the pressure increases. Uranium nuclei decay to lighter nuclei but the density remains constant. Macroscopic pieces of uranium fly away due to high pressure but this process is not determined by the laws of General Relativity. Thus, different laws like chemical, nuclear or particle are not derived from the General Relativity. They are additional laws ruling interaction of different kinds of particles.Nevertheless the idea of one fundamental field describing everything maybe is not dead. It might happen that a unified geometrical theory will revive despite the large number of different particles and fields which lead to considerable difficulties in its realization.

3.4   Massless and massive fields - long range forces

Although there are many differences between electrodynamics and gravitation they have one thing in common. Both of them are long range interactions. The forces fall off as r^{-2} at large distance (and the corresponding potentials fall as r^{-1}). The existence of atoms, of planetary systems and ultimately of life is possible thanks to this property. It is connected with the fact that quanta of these fields (photons and gravitons) are massless. Indeed the solution of equation (3.3) which describes massless field is $\phi \sim r^{-1}$. For $m \neq 0$ (see eq. (3.4)) the solution turns into:

$$\phi \sim \exp(-mr)/r \tag{3.22}$$

The field is negligibly small when $r > m^{-1}$.

Quantum field theory claims that vanishing of the mass is a non-trivial property of a field. Very specific conditions must be imposed on the theory to guarantee that the field would not acquire mass due to quantum corrections. If no such conditions are fulfilled then the primarily massless field emitting and absorbing quanta of other fields necessarily becomes massive.

Both gravitational theory and electromagnetism have the property that forbids mass (i.e. A^2m^2 terms in Lagrangian) not only on a classical level but also on the quantum one. These properties are gauge invariance in electrodynamics and general covariance in gravitation. There is a deep connection between these principles and electric current conservation in electromagnetism or energy momentum tensor conservation in gravitation. These conservation laws are implied by the above mentioned principles. The reverse statements are not true however. A theory of a massive vector field that interacts with a conserved current can be constructed.

The natural question arises whether the gravitational and electromagnetic forces are the only long range ones in Nature or whether there exist other as yet undiscovered fields and corresponding particles?

Those who are acquainted with cosmology know that there are theories assuming the existence of one more massless, in this case scalar, field ϕ. These are the so called scalar-tensor gravitational theories (e.g. Brans-Dicke theory) in which gravitational field is described by a mixture of scalar and tensor fields. As it was mentioned above the scalar field must acquire mass due to quantum effects. The potential of this field becomes short range and the influence of it at large distances becomes negligible. The estimates of the field mass vary from $m = m_{Pl}$ with corresponding interaction radius $l = 10^{-34}$cm, to $m = m_0^2/m_{Pl}$ where m_0 is a characteristic mass inherent to the model. If m_0 is connected with supersymmetry breaking (see ch.4) i.e. $m_0 = 10^2 - 10^3 \text{GeV}$ then $l = 0.1 - 10\text{cm}$; if it is connected with strong interactions then $m_0 = 0.1\text{GeV}$ and $l = 10^7 \text{cm}$. Such a field is unobservable on astronomical scales. Hence the astronomical tests of Brans-Dicke and similar theories give negative results.The existence of such a field ϕ could be proved by the discovery of the variation of the gravitational constant on a scale $l = m^{-1}$.

No other massless tensor field except for gravitational, is permitted because the source of such a field must be a conserved second rank tensor. The only such tensor is the energy-momentum tensor (there exists a special theorem in field theory which proves it).

There could be many more massless vector fields. A lot of conserved vector currents can be constructed in addition to the electromagnetic current. The question is if there are long range interactions connected to these currents. Thirty years ago T.D.Lee and C.N.Yang asked if there existed a long range interaction coupled to baryonic charge B. This is the charge prescribed to protons, neutrons and some other particles in order to decribe the conservation of their number observed in experiments (see ch.4).

If there were a long range interaction coupled to baryonic charge then the number of baryons in a body could be measured indirectly using the Gauss theorem.

It has, however, been proved that even if such an interaction existed, it would be negligibly small. The proof is based on the experimental accuracy of the equivalence principle. According to this principle gravitational mass is equal to inertial one, i.e. gravitational acceleration is the same for all bodies. This has been checked by V.B.Braginsky and V.I.Panov with 10^{-12} accuracy with respect to the attraction to the Sun. If there were long range forces between baryonic charges then these accelerations would not be equal. The point is that the mass of a nucleus is not exactly proportional to its baryon number. The deviation is about a few thousandths. The negative results of these experiments show that a baryonic field, if it exists, has a coupling which is $10^{-3} : 10^{-12} = 10^9$ times weaker than the gravitational coupling and 10^{45} weaker than electromagnetic one. This unnatural smallness leads one to believe that such interactions do not exist at all. It should be noted, however, that very small numbers may appear in physics. We know that in quantum field theory factors as small as $\exp(-1/\alpha)$ can arise where $\alpha \approx 10^{-2}$ is the appropriate coupling constant. Thus the existence of any interaction can never be ruled out but can only be experimentally bounded.

If a vector field had a nonvanishing mass then the interaction potential would be of the type of eq.(3.22) instead of $U \approx r^{-1}$. The bounds on strength of such an interaction depend on m and are very weak if m is large.

Thus if there is no long range force coupled to baryonic charge it can be assumed that the latter is not conserved. As it was already mentioned it is necessary for the explanation of the baryon asymmetry of the Universe.

In conclusion we would like to stress that although the long-range character of electromagnetic and gravitational interactions has a good theoretical basis, the last word belongs to experiment. The inverse square law in gravitation is not checked on galactic scales. Thus it is tempting to assume that the phenomena usually ascribed to dynamical hidden mass can be explained by a modification of the gravitational interaction. The considerations of this section show that this can hardly be done by introducing additional massless fields. There is no difficulty from the theoretical nor from the experimental point of view for modifying gravity at small scales $(r = m_{Pl}^{-1})$, but at large scales one would need to change the existing theory in a non-trivial way.

Chapter 4

Elementary particles

This chapter contains a short introduction to elementary particle physics. Its aim is to present to a nonspecialist (e.g. to an astronomer) the fundamentals of the theory and explanation of basic terms, so that one can better understand the following chapters. Three books by Okun (1985, 1983, 1981) are recommended as supplementary reading. The first one contains a popular introduction to particle theory digestible even for high school pupil while the last is of interest even for professionals.

4.1   Scalar fields and π mesons

Though the scalar field proved to be useless in gravity it found an application in physics as the field responsible for nuclear forces. It was proposed in 1937 by Yukawa. He wrote the field equations:

$$\partial^2 \phi = 4\pi g n_{bar} + \mu^2 \phi \tag{4.1}$$

where g is the coupling constant of nuclear interaction. The source of the field n_{bar} is a scalar quantity. It is equal to baryon number density in nonrelativistic limit. Because of last term $\mu^2 \phi$ the field quanta are massive and nuclear forces are short range. A static, spherically-symmetric solution of equation (4.1) outside the source is:

$$\phi = const \times e^{-\mu r}/r$$

The potential falls sharply at $r > \mu^{-1}$ due to the exponential factor and the nuclear forces are negligible at $r = (2 - 3)\mu^{-1}$. In order to be in agreement with experiments the value of μ had to be chosen as the inverse of the nuclei size 10^{-13}cm that is $\mu \approx 100$ MeV.

Equation (4.1) also has wavelike solutions like electromagnetic waves:

$$\phi = \phi_0 \exp(i\omega t - ikr)$$

The relation between frequency ω and wave vector k is called dispersion relation. The dispersion relation for massless fields, for example the electromagnetic one, is:

$$\omega^2 = k^2 \tag{4.2}$$

This means that the frequency can be arbitrarily small. The dispersion relation for a massive field is:

$$\omega^2 = k^2 + \mu^2 \tag{4.3}$$

The frequency cannot be smaller than μ^2.

Velocity of energy transport by this field which is in fact the group velocity is equal to:

$$V_{gr} = \frac{k}{\sqrt{k^2 + \mu^2}}$$

and changes from zero (for $k = 0$) to the speed of light (for $k \to \infty$). Let us note that whereas wave packets with dispersion relation (4.2) do not change their form, wave packets with dispersion relation (4.3) deform when moving.

Particles with the mass predicted by Yukawa were soon found. They were called mu mesons and their mass was 105 MeV. It appeared soon, to everybody's surprise, that they did not practically interact with nuclei. Later it was found that these particles are not scalar but have spin $1/2$. Today they are called muons and the term "meson" is used for another group of particles.

Soon however scalar particles with mass 135 MeV which strongly interact with nuclei were found. They were called π mesons.

It appeared later that there were three kinds of π mesons: two charged π^+, and π^-, and one neutral, π^0. These particles carry nuclear forces. The golden age of scalar field theory came after this discovery. Isotopic invariance was found and three π mesons were described as three different states of one field. The proton and neutron can also be described as one nucleon field exactly like spin-up-electron and spin-down-electron are the same particle in two different states.

In analogy with usual spin, different states of nucleon and pion fields are characterized by so called isotopic spin (or shortly isospin). The proton corresponds to an isospin up state while the neutron corresponds to an isospin down state, both being components of a spin $1/2$ isotopic spinor. π mesons form an isotopic vector i.e. their isotopic spin is 1. Exactly in the same way as physical laws do not depend on the rotations of coordinate system, the nuclear forces are independent from rotations in internal isotopic space. This property is called isotopic invariance. Isotopic rotations make the transformations $p \leftrightarrow n$, $\pi^+ \leftrightarrow \pi^0 \leftrightarrow \pi^-$. Different electric charges lead to different electromagnetic interactions of these particles but they are small in comparison with their nuclear interactions.

To be more exact let us note that π meson field is not scalar but pseudoscalar. Space reflection changes its sign.

4.2 Quarks and the structure of hadrons

It became known later that the theory of nuclear forces, based on π mesons exchange and isotopic invariance, is not always applicable and the π meson field is a composite one. Hence the notion of the scalar field as a fundamental object was abandoned [1]. Vector fields took its place. Experiments showed that baryons (protons, neutrons), as well as mesons, are composite particles consisting of more elementary ones, called quarks. The latter are fermions and their spin is 1/2. Only two quarks u and d are needed to describe nucleons and π mesons. The electric charge of quarks is smaller than that of the electron (-e):

$$q_u = \frac{2}{3}e, \quad q_d = -\frac{1}{3}e$$

There is a large class of particles called hadrons which are built of quarks. Among them are nucleons, π mesons and many others. Leptons i.e. electron, muon, neutrino are not built of quarks. The condition which determines the quark structure of hadrons is that the net electric charge of a hadron should be integer: p=(uud) (proton consists of two u quarks and one d quark), n=(udd), $\pi^+ = (u\bar{d})$, $\pi^- = (\bar{u}d)$, $\pi^0 = (u\bar{u} - d\bar{d})/\sqrt{2}$.

Quarks have one more important quantum number which is called colour. It was introduced due to the following reason.

In reactions of π meson scattering on nucleons a resonant intermediate state was found, which was called the Δ-resonance:

$$p + \pi^+ \rightarrow \Delta^{++} \rightarrow p + \pi^+$$

(the isotopical partners of Δ^{++} such as Δ^+, Δ^0 and Δ^- were also discovered). A resonance in particle interactions is a phenomenon very similar to resonant light absorption in matter. When light passes through matter and the light frequency is close to that of an atomic transition the light absorption is very strong. Atoms become excited due to the light absorption. Excited atoms decay emitting either a few low energy photons or one photon of the same energy. In both cases there exists an intermediate state, the excited atom. Such excited atomic state is analogous to the excited quark state Δ.

---

[1]It will be shown later that the scalar field soon returned to particle physics and then appeared in cosmology.

The mass of Δ^{++} was found by the measured energies of p and π^+. It is 1240MeV. The charge of Δ^{++} is twice the charge of proton and the spin is 3/2. Thus Δ^{++} consists of three u quarks, all their spins being parallel. According to the Pauli principle, no more than one fermion can be in one quantum state. Therefore the quarks should somehow differ from each other. The simplest solution, that the difference lies in the quark angular momentum leads to a wave function antisymmetric in quark coordinates. Such wave function does not describe low energy parameters of Δ^{++}. Less trivial and as it proves more fruitful explanation is that quarks are not identical. To distinguish between quarks the notion of quark colour was proposed. Each quark can exist in three different colour states i.e there are following quarks: u_r, u_b, u_y, d_r, d_b, d_y (the indices being red, blue and yellow). Thus Δ^{++} can be described as the ground state of three quarks of different colours u_r, u_b, u_y.

4.3 Quark interactions - quantum chromodynamics

The discovery of quark colour was the first step to strong (nuclear) interaction theory. The next step was the hypothesis that colour is a charge with which vector fields called gluons (g) interact [2]. This interaction is similar to that of photon with electric charge. The important point is that quark colour can be changed by interaction with a gluon:

$$u_r \rightarrow u_b + g_{rb}$$

The idea of a particle property changing in an interaction comes from quantum electrodynamics. Scattering of electrons can be described as a two step process. First:

---

[2]This theory resembles the Yukawa theory with the formal difference that instead of (pseudo)scalar meson exchange between nucleons vector gluon exchange between quarks is responsible for the interaction. This point, not very essential at first sight, leads to a deep qualitative difference between the theories.

$$e \to e' + \gamma$$

i.e. electron e "decays" into electron e' with another momentum and polarization and photon γ (virtual photon). Then

$$\gamma + e'' \to e'''$$

that is electron e'' absorbs this photon, and electron e''' is created. Although in electrodynamics only polarization and momentum of an electron are changed this is where the idea of change of particle type comes from. Gluon emission or absorption changes the quark colour while other internal quantum numbers of quarks (e.g. electric charge) remain the same. That is a u-quark can be transformed only into a u-quark of the same or different colour and similarly for a d-quark, etc. It is said that a gluon changes the quark colour but does not change its flavour. Gluons have no electric charge but they do have colour (from the point of view of interaction gluons are like charged photons). If they were colorless and still a quark could change its colour by gluon emission then colour would not be conserved. The fact that gluons have colour leads to their self interactions and nonlinearity of their equations of motion. (Let us recall that General Relativity equations are non-linear since gravitons have energy and themselves gravitate).

There are many gluons with different colour charges while there is only one photon. Gluon g_{ij} changes quark of colour i into quark of colour j. The matrix g_{ij} has $3 \times 3 = 9$ elements but the identity operator $\delta_i^j \, \mathrm{tr} g$ is subtracted as it does nothing to quarks. Thus there are 8 gluons.

Quark interactions can be described by gluon field lines in the same way as electromagnetic interactions are described by electromagnetic field lines. There is, however, a very important difference. When two electrons are pulled apart the electric field lines become radial near each electron. The electrostatic energy decreases as $1/r$ and correspondingly the force goes down as $1/r^2$. The gluon field contrary to the electromagnetic one has (colour) charge with which it interacts itself i.e. the field is essentially non-linear. It is believed now that because of this phenomenon the gluon field lines

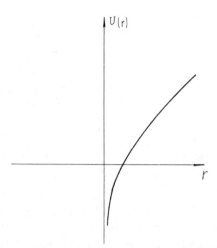

Figure 4.1: Quark interaction potential as a function of distance.

do not deviate when quarks are pulled apart but form a string-like configuration.

The energy of such a configuration is proportional to the distance between the quarks i.e to the string length. Thus the quark interaction potential is Coulomb-like at small distances and grows linearly at big distances (Fig. 4.1). This leads to the new phenomenon of quark confinement. Quarks, contrary to protons, electrons etc., cannot exist as free particles. In attempts to pull quarks far away from each other by e.g. crushing protons in proton-proton collisions, a quark may receive a large portion of energy, and draw the gluon string. The gluon field becomes very strong so that it can produce quark-antiquark pairs from the vacuum. The energetic quark together with a newly formed antiquark may escape forming e.g. a π meson.The same happens in attempts to liberate a gluon. Generally, only colorless (white) states can escape to macroscopic distances. All the known elementary particles are white. Let us note that not only quark-antiquark pairs, but also three quarks can form a colorless state e.g. proton or neutron. Here lies a very important difference between chromodynamics and electrodynamics.

All colorless states have integer electric charges and, vice versa, all quark combinations with fractional electric charge must be colored. Coloured states with integer electric charge are also possible but they cannot exist in free form.

The colour confinement has not been proved from first principles but the presented picture is well confirmed by experiments. In particular, hadronic jets should appear when gluon strings are broken. These jets have been observed in many experiments. A hadronic jet is a quark trace in vacuum similar to an electron trace in a Wilson chamber.

Quark interactions at small distances ($r < 10^{-13}$ cm) are very simple from a theoretical point of view. They are like electromagnetic interactions. There exist quarks which are much heavier than u and d, and their bound states, observed in experiments, resemble positronium. Now the existence of five quark types (flavours) is definitely established by experiment and the sixth one is expected on theoretical grounds.

q	u	d	s	c	b	t
m_q	$5MeV$	$7MeV$	$150MeV$	$1.2GeV$	$5GeV$	$\sim 100GeV$?

The proton mass is 940 MeV. This is much larger than the sum of the constituent quark masses. The energy of the gluon field and the kinetic energy of quarks give an essential contribution to it. The values shown above refer to bare quark mass when it is measured with large momentum transfer (i.e. at small distance) when the gluon field contribution is small. A potential approach is inapplicable to u, d and s-quarks because they are relativistic inside a particle. However, the bound states of new heavy quarks, their energy levels and transition probabilities are well described by the potential of the form (fig. 4.1):

$$U = \frac{\alpha_c}{r} + br$$

Many particles consisting of $b\bar{b}$ and $c\bar{c}$ pairs have been discovered. The agreement between the theory and experiment confirms the idea of quark structure of hadrons.

Chromodynamics becomes very sophisticated at large distances ($r \approx 10^{-13}$ cm) because the effective colour coupling constant, describing quark-gluon interactions, grows with distance and is about

1 at $r \approx 10^{-13}$ cm, whereas it is small at small distances or at large momentum transfer. This phenomenon is called asymptotic freedom. The large effective charge causes many interesting phenomena at $r \approx 10^{-13}$ cm and leads to very complicated hadron world.

Nuclear forces are understood to result from a screened colour interaction. They are analogous to the electromagnetic van der Waals forces between neutral atoms. These forces are relatively weak. They lead to a nuclear binding energy of about 10 MeV whilst the characteristic energy of the quark interaction is an order of magnitude (or even two orders) larger.

Nuclear forces cannot be described in the framework of perturbation theory, because the colour charge is not small at large distances. Still the long distance part of this interaction connected with the exchange of a π meson (which is the lightest hadron) can be described in a rather simple way.

Of course the notion of nuclei consisting of protons and neutrons has not been altered by the quark model. Quarks are not essential in nuclei in zero order approximation exactly like nuclear structure is not important in chemistry. However, the quark structure of hadrons is important in a more precise description of nuclear phenomena.

It is not clear whether quarks have internal structure or not. Regardless of the answer, the quark model will still be a good approximation to physical phenomena in the energy range from 1 to 100GeV.

As we have already mentioned six kinds of quarks are known today (5 for certain). Three of them : u, c and t have electric charge $+\frac{2}{3}e$ and the other three d, s and b have charge $-\frac{1}{3}e$. Each quark has three colour states, which makes 18 types of quarks. Neither mass, nor electric charge depend on colour so one says that there are 6 kinds of quarks or 6 quark flavours. The gluon does not feel the quark flavour, just like the photon cannot distinguish between muon and electron.

4.4   Quark lepton symmetry - a bound to the number of species of leptons and quarks

There is a curious symmetry between quarks and leptons in nature. Each pair of quarks corresponds to a pair of leptons:

$$\begin{pmatrix} u \\ d \end{pmatrix} \leftrightarrow \begin{pmatrix} \nu_e \\ e \end{pmatrix}, \quad \begin{pmatrix} c \\ s \end{pmatrix} \leftrightarrow \begin{pmatrix} \nu_\mu \\ \mu \end{pmatrix}, \quad \begin{pmatrix} t \\ b \end{pmatrix} \leftrightarrow \begin{pmatrix} \nu_\tau \\ \tau \end{pmatrix}$$

This quark-lepton symmetry is probably not accidental. In the unified theory of electroweak interactions this is the sufficient condition for the cancellation of the so-called axial vector anomaly. Otherwise the theory would not be renormalizable.

Only three quark-lepton families are known experimentally now. There could exist other families which have not been observed yet because of the insufficient energies of existing accelerators. The quark-lepton symmetry permits us to obtain an upper bound to the number of quark flavours. This bound is based on cosmological arguments (chap.1. sec.8.). Each neutrino corresponding to a pair of quarks is contained in the primordial plasma during nucleosynthesis (1-100s from the beginning). The number density of more heavy quarks and charged leptons, except for electrons, at this time is negligible. Each neutrino species contributes to the energy density and thus changes the speed of the expansion and of the cooling down of the Universe. On the other hand, the expansion rate of the Universe has impact on the output of light nuclei. Taking the central values of the observed abundances of He^4 and H^2 one can conclude that $k_n < 4$ and no new quark flavours exist. All quark flavours are already known !

The validity of this conclusion will be checked experimentally in the near future by the measurement of the lifetime of the neutral intermediate boson Z^0 which depends on the number of neutrino species.

4.5 Electroweak interactions - intermediate bosons

Another example of a vector field in physics is the field of the $W^\pm$ and Z^0 bosons (they are sometimes called wions). They are carriers of weak interactions. The first encounter with these interactions was nuclear β decay which is the result of neutron decay:

$$n \to p + e^- + \bar{\nu}_e$$

Let us note that though the free neutron is not stable it can become stable inside a nucleus if the difference between binding energies of the decayed neutron and the produced proton is greater than the energy released in the decay. Because of that stable nuclei can exist. If the neutron binding energy in a nucleus is large and that of the proton is small then an inverse process may proceed:

$$p \to n + e^+ + \nu_e$$

The neutron lifetime is about 15 minutes [3], which is much larger than one would expect from the mass defect estimate:

$$t \approx 1/\Delta m \approx 10^{-21} s$$

This proves that the interaction responsible for the decay is very weak and explains its name.

As there are four fermions taking part in this interaction, physicists started to call it a 4-fermion interaction. About 50 years ago Fermi wrote the amplitude of the decay in the following way:

$$g_F \Psi_p^* \Psi_n \Psi_e^* \Psi_\nu \tag{4.4}$$

where Ψ_p is proton wave function, Ψ_n is neutron wave function etc., g_F is the weak interaction coupling constant.

---

[3] Lifetime is by definition the period of time during which the number of decaying particles becomes $e = 2.71...$ times smaller. Sometimes a decay is characterized by its so-called half-life $\tau_{1/2}$. This is the period during which the number of particles becomes 2 times smaller. For the neutron $\tau_{1/2} \approx 10 min$.

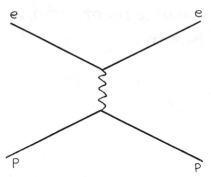

Figure 4.2: The diagram describing the electron-proton scattering. The waved line corresponds to exchanged photon.

Not only neutron decay is described by expression (4.4) but according to quantum field theory, so are the processes:

$$n + e^+ \leftrightarrow p + \bar{\nu}_e$$

$$n + \nu_e \leftrightarrow p + e^-$$

and so on. Let us note that in these reactions the number of baryons (baryonic charge) as well as the number of leptons (leptonic charge) and, of course, electric charge are conserved.

An important property of equation (4.4) is that all four wave functions are taken in the same spacetime point. This is called a local interaction. A very similar electromagnetic process on the other hand:

$$e + p \rightarrow e + p$$

is not local. It can be described by the diagram of Fig. 4.2. A proton emits a virtual photon at point x. This photon is absorbed by an electron at point y. The amplitude of this process includes the product of $\Psi_e(y)\Psi_{e'}(y)$ and $\Psi_p(x)\Psi_{p'}(x)$. Local amplitude (4.4) is illustrated by the diagram presented in Fig. 4.3.

The success of electrodynamics on the one hand, and the difficulties in describing the four-fermion interaction connected with its non renormalizability on the other, have led to attempts to formulate weak interaction theory along the same lines as electromag-

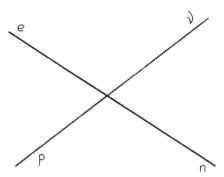

Figure 4.3: The diagram describing the process $e^- + p \rightarrow n + \nu$. It can be regarded as a local process when the energy is low.

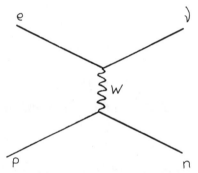

Figure 4.4: $e^- + p \rightarrow n + \nu$ scattering in the electroweak theory. The nucleon and lepton currents interact nonlocally through W-boson exchange.

netism. Another reason to build such a theory was the similarity of electromagnetic and weak currents. In both cases the amplitude can be written as a product of two vector currents; $J^\mu J_\mu$, where J_μ is a product of two fermion wave functions. Moreover, the weak vector current is conserved just like the electromagnetic one. Hence, there appeared a natural hypothesis that weak interactions also proceed through an exchange of a vector particle (Fig. 4.4).

This particle called a W-boson was observed and its mass is 80GeV in agreement with the theory. The large mass ensures that weak interactions are approximately local, their interaction radius is 10^{-16} cm. The weak and electromagnetic interactions are very similar at distances smaller than 10^{-16} cm. Electromagnetic forces are long range while weak forces fall sharply at $r > 10^{-16}$ cm since W-boson is massive. Another difference is that W-bosons are charged, and photons are neutral.

The large mass of a W boson makes the process $n \rightarrow p + W^-$ energetically impossible. It does not mean that such processes are forbidden and the neutron decay scheme is not true. The point is that according to the uncertainty principle in quantum mechanics energy is conserved only for a large time interval. During a period Δt energy is conserved only with accuracy of $h/\Delta t$. A neutron may for a short time turn into proton and W^- boson. This is a so-called virtual process and the W^- is a virtual particle. Knowing the W^- mass, the lifetime of this virtual state can easily be calculated:

$$t_W = 1/m_W = 10^{-26}s$$

Accordingly, the size of the interaction region is:

$$l = t_W = 10^{-16}\text{cm}$$

In the formal language of quantum mechanics the process:

$$n \rightarrow p + W^-$$

is described as an admixture of $(p + W^-)$ state to neutron state.

The notion of mixed state appeared long before weak interaction theory. There are S- and P-states in the hydrogen atom. Putting this atom in electric field E one gets a mixed state:

$$\Psi = \Psi_S + c\Psi_P$$

Amplitude c is proportional to:

$$c \sim \frac{eEd}{E_s - E_p} \qquad (4.5)$$

where $(E_s - E_p)$ is the energy difference between P and S states and d is the "size" of hydrogen atom.

Quite analogously the transition $n \leftrightarrow p + W^-$ inherent in neutron decay is described by admixture of $(p + W^-)$ to neutron with the amplitude $c_1 = (e/\sqrt{hc})\Delta m/(m_W + m_p - m_n)$. The denominator is the difference between energies of the initial and the final states, which usually appears in the perturbative calculations. The factor Δm in the numerator is connected with the amplitude of the process $n \to p + W^-$. As this process is energetically impossible, the second order perturbation should be considered i.e. the transition $W^- \to e + \bar{\nu}_e$. This transition is described by admixture of $(e + \bar{\nu}_e + p)$ to $(p + W^-)$ with the amplitude:

$$c_2 = \frac{e}{\sqrt{\hbar c}} \frac{\Delta m}{m_W + m_p - m_p - (E_e + E_{\bar{\nu}_e})/c^2} \qquad (4.6)$$

The admixture of $(e + \bar{\nu}_e + p)$ to initial neutron is $c_1 c_2$. The probability of transition per unit time can be estimated as:

$$P = |c_1 c_2|^2 \frac{\Delta m c^2}{\hbar} \approx \frac{\Delta m c^2}{\hbar} \left(\frac{\Delta m}{m_W}\right)^4 \left(\frac{e^2}{\hbar c}\right)^2 \qquad (4.7)$$

The coupling constant of weak interactions has been assumed to be close to the electromagnetic one because W is very much like photon. This proves to be true in exact theory. The result contains two small factors $(\Delta m/m_W)^4$ and the square of the fine structure constant. They diminish the naive estimate $10^{21} s^{-1}$ of the decay rate to the observed $10^{-3} s^{-1}$.

The theory of weak interactions realized by intermediate boson exchange is based in its final version on works by Weinberg(1967), Salam (1967) and earlier works by Glashow. Apart from $W^{\pm}$ bosons there exists a neutral Z^0 boson in this theory. Its mass is slightly larger than that of W. This particle has been observed and its mass was measured to be 90 GeV. Z^0 resembles a heavy photon but it interacts not only with charged particles, but also with neutral ones.

W bosons are responsible for the following transitions:

$$n \to p + W^-, \quad e^- \to \nu_e + W^-, \quad \mu^- \to \nu_\mu + W^-, \text{etc.}$$

while Z^0 leads to charge conserving transitions:

$$p \to p + Z^0, \quad n \to n + Z^0, \quad e^+ \to e^+ + Z^0, \nu \to \nu + Z^0, \text{etc.}$$

and as a result of them to the reactions:

$$N + \nu \to N + \nu, \quad e^+ + \nu \to e^+ + \nu, \text{etc.}$$

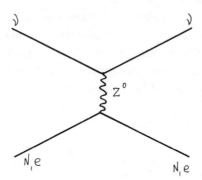

Figure 4.5: $\nu N \rightarrow \nu N$ or $\nu e \rightarrow \nu e$ scattering due to weak neutral current (Z^0 boson exchange).

which are described by the diagram of Fig. 4.5. These are so called neutral current processes. In particular there exists elastic neutrino scattering on protons and electrons.

These processes are very important in astrophysics. Before the modern weak interaction theory was known it was believed that energy radiates from the inner parts of the stars because of the URKA processes. Neutrinos born in these processes could easily escape from the star thus cooling it down. We know now that neutral currents should be taken into account. The cross section of neutrino scattering has become larger and stars cool down slower.

The predictions of the new theory are interesting not only for astrophysics but also for cosmology. If only charged currents existed electron-positron annihilation to neutrinos would proceed through W exchange producing only electron type neutrinos (Fig. 4.6). The existence of Z^0 allows also the reactions:

$$e^+ + e^- \rightarrow \nu_\mu + \bar{\nu}_\mu, \quad \nu_\tau + \bar{\nu}_\tau$$

which are illustrated by figure 4.7. These processes maintain the equilibrium of ν_μ and ν_τ without heavy μ and superheavy τ. Hence, all three types of neutrinos were in thermal equilibrium down to temperatures of about 5MeV, which is very important for primordial nucleosynthesis.

There is one more difference, not yet discussed, between weak and electromagnetic interactions. Weak interactions break parity.

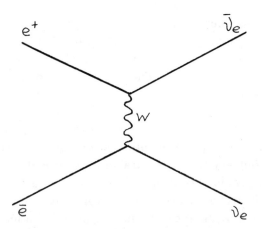

Figure 4.6: Process $e^+ + e^- \rightarrow \nu_e + \overline{\nu_e}$ due to charged weak currents ($W^\pm$ boson exchange). Electron interacts with electron neutrino only.

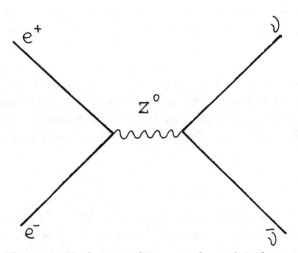

Figure 4.7: The interaction between electrons and neutrinos due to neutral weak currents (Z^0 exchange). Electrons interact with every type of neutrinos.

In particular it is manifested in the following. The spin of mass-less particles is either parallel to their velocity or antiparallel. Both photon polarizations are possible. The ratio between left and right polarization photons born in reactions with nonpolarized particles is one to one. At equal frequencies different polarization states of photons are degenerate and one can consider any superposition of these states and in particular a sum and/or difference of them corresponding to linear polarization. There is only one polarization state of the neutrino in which the spin and momentum are antiparallel; and on the contrary they are parallel for antineutrinos.

A classical analogy is a corkscrew. A corkscrew can move ahead in a cork only when turned around from left to right. There are no corkscrews moving ahead turned from right to left. That is the parity breaking. Parity would be conserved if there were half of left to right corkscrews and half right to left corkscrews.

Parity breaking has nothing to do with the vanishing mass of the neutrino. If $m_\nu \neq 0$ polarization would depend on the coordinate system. In fact there is always a coordinate system moving faster than a massive neutrino. Momentum has the opposite direction in this system while spin remains the same. Thus, momentum and spin are parallel and the neutrino is right-handed. Both polarization states are possible no matter how the neutrinos are produced. In this case the parity breaking is manifested by a different amount of left-handed and right-handed neutrinos produced in one or other reaction. If $m_\nu = 0$ then only left-handed neutrinos exist and if $m_\nu \neq 0$ the relative number of right-handed neutrinos is proportional to:

$$(1 - v) \approx (m/E)^2$$

This law was checked experimentally for electrons. Electrons produced in β decay are mostly left-handed. The degree of their polarization is v/c. It means that W-bosons interact only with one (left) polarization state of electron, proton, neutrino etc. This was measured in the experiment in which parity violation has been proved.

Beta decay of radioactive cobalt Co^{60} was chosen to test whether parity was conserved. The sample of Co^{60} was cooled down to the temperature of liquid helium in order to minimalize heat fluctuations. Nuclear angular momenta were oriented by magnetic field. The polarized nuclei decayed as follows:

$$Co^{60} \rightarrow Ni^{60} + e^- + \bar{\nu}_e$$

The angular distribution of electrons relative to the angular momentum of Co^{60} was measured. Let us assume for definiteness that the angular momentum of Co^{60} was directed upward. If parity were conserved then equal number of electrons would be observed in up and down hemispheres. The experiment showed,however, a large asymmetry. The electrons were predominantly emitted into the lower hemisphere opposite to the nuclear spin direction. The mirror image of that is the process when electrons mostly go in the nucleon spin direction. This process is physically impossible. Hence parity is not conserved.

Weak interaction theory is specially formulated so that W - bosons interact only with left-handed particles and right- handed antiparticles. A theory in which Z^0 bosons interact in the same way could be formulated but it was more attractive to unite weak and electromagnetic interactions in a single theory. Photons do not distinguish one particle polarization from another. Because of that Z^0 bosons also interact with both left-handed and right-handed particles but with each in a different way. Thus Z^0 breaks parity in contrast to photons. In particular Z^0 would not interact with right-handed neutrinos if they existed. The difference between quarks and charged leptons, which have both polarization states, and neutrinos with one polarization state, should be stressed. This difference persists even if neutrinos are massive and have the so called Majorana mass which is permitted for a neutral fermion. This mass leads to transition of a left handed particle into a right- handed antiparticle, whereas more common Dirac mass leads to a transition of left-handed particle into right-handed particle.

In electroweak interaction theory the photon, Z^0, W^+, W^- are considered as manifestations of one vector field. More precisely, there are two vector fields: neutral B_Y^0 interacting with weak hypercharge Y which we will define later on, and a triplet (B_I^0, W^+, W^-) interacting with so-called weak isospin I. This last field is analogous to π meson triplet (π^+, π^0, π^-). Each pair of leptons is considered as one lepton field just like proton and neutron are considered as a single nucleon field. Lepton pairs (e, ν_e), (μ, ν_μ) and (τ, ν_τ) are weak isospin doublets, and so are quark pairs of the same colour e.g. (u_b, d_b) etc. Weak hypercharge is defined as twice the mean charge in a multiplet $Y = 2(Q - T_3)$. Weak dublets are formed by left-handed particles only and therefore (W^+, W^-, B_I) interact with left-handed particles. Right-handed particles form isotopic singlets with hypercharge $Y = 2Q$. Therefore the field B_Y interacts with both, left- and right-handed particles. Because of spontaneous symmetry breaking (see sec.8 of this chapter) a linear combination of B_Y and B_I acquires mass and is identified with Z^0 while the orthogonal combination (that is the photon) remains massless.

Let us note that electroweak interactions do not lead to confinement as chromodynamics does. This difference is caused by symmetry breaking in weak interactions. Roughly speaking, nonzero masses of W and Z save them from confinement. Massless photon is not confined since it is not charged and there is no photonic self-interaction. In other words the abelian gauge symmetry $U(1)$ does not lead to confinement.

Now we would like to say a few words about neutrino oscillations. If the neutrino mass is not equal to zero, then the states which form weak doublets may not have a definite mass, for example $\nu_e = c_1 \nu_1 + c_2 \nu_2 + c_3 \nu_3$ where ν_1, ν_2 and ν_3 have masses m_1, m_2 and m_3 respectively. In this case ν_e produced, say, in the reaction $ep \to n\nu_e$ transforms into ν_τ and ν_μ and back when it propagates in vacuum, and the probability of electron production in the inverse process $\nu_e n \to pe$ is an oscillating function of distance. This might explain the lack of solar neutrinos in the Davis experiment.

The above mentioned parity violation manifests itself not only in particle reactions, but also in ordinary matter though very weakly. Let us consider for example a hydrogen atom.

An electron is bound in an atom by electromagnetic forces. There is, however, an additional proton-electron interaction mediated by the Z^0 boson. It is very difficult to observe this interaction by the shift of the energy levels as it gives a small correction to a big value. It can be discovered, however, thanks to parity violation which is caused by Z^0 exchange. Because of this interaction any atom, not only hydrogen one, develops small parity violating effect in interactions with light. In particular, it manifests itself in rotation of light polarization by matter.

The effect is larger for heavy atoms because for them the probability of finding an electron close to the nucleus is greater and so is the probability of Z^0 exchange. This tiny effect was discovered in the Novosibirsk Nuclear Institute in 1979. Thus it was established that parity is violated in atomic physics.

Another experiment which has confirmed parity violation in electron-proton interactions was done in Stanford in 1981. The difference between the cross sections of scattering of left- handed and right-handed electrons on deuterium was measured there. The results are in excellent agreement with Glashow-Weinberg-Salam theory.

The intermediate bosons of weak interactions $W^\pm$ and Z^0 were first observed only recently in 1984 in the CERN proton-antiproton collider. The particles cannot be observed directly because their lifetime is very short. It is about 10^{-24} s. Their path during this time is 10^{-14} cm. It is smaller than the electron diameter. Their existence can be proved by observing the products of their decay. Their masses can be calculated from the energies of the decay products. More than 10,000 such bosons have already been observed.

4.6 Grand unification theories and baryon non-conservation

The unification of weak and electromagnetic interactions stimulated the next step: the search of a unified theory of strong and electroweak interactions. Many models have been built in which gluons, as well as weak intermediate bosons, were described as different manifestations of one fundamental vector field. Quarks and leptons were described as manifestations of one fermion field. All

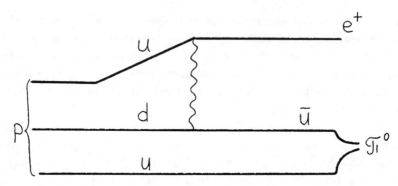

Figure 4.8: The diagram describing proton decay. The waved line corresponds to intermediate bosons of grand unification that transform quarks into leptons.

theories demanded that in addition to the 12 already known vector bosons there must exist an equal or even larger number of new particles called X- and Y-bosons. Just like gluons change colour of a quark, X and Y bosons change charge of a quark or a lepton and moreover can transform quark into lepton. That means that baryon number is not conserved in interactions with these bosons. In particular, they would lead to proton decay illustrated by the diagram of figure 4.8.

The theory predicts very high masses of X and Y bosons - about 10^{15} GeV. That is the reason why, despite intensive searches, no proton decay has been observed yet. The situation is rather disturbing now because the experimental limits on proton lifetime are higher than predictions of the simple unification models. Today the only experimental argument in favour of baryon non-conservation is based on cosmology. It is the baryon asymmetry of the Universe.

With the appearance of particles as heavy as 10^{15} GeV cosmology provides us with the only way except for the proton decay to learn something about physics at such energy scales.

The existing lower limit on the proton lifetime is about 10^{32} years. In the coming years it may be raised by one or even two orders of magnitude. After that the laboratory possibilities to prove baryon non-conservation will be exhausted. In this connection it is noteworthy that the inflationary model definitely indicates baryon non-conservation after inflation (see section 3 ch.8).

The mechanism of baryon non-conservation discussed here is not the only one. Other possibilities are discussed in chapter 8.

4.7 Higher orders of perturbation theory and renormalization

The reader might think that all the interactions except for the gravitational one are described by vector fields, and that scalar fields are of no interest. In the recent years, however, the general attitude towards theory has changed altogether. If earlier it was considered that anything which is not proved does not exist, now everything which is not forbidden is allowed. The last principle should be enough to consider the possible role of a scalar field (not found experimentally) in cosmology. Nevertheless, scalar fields became popular only with the development of unified theories of particle interactions(like electroweak interaction theory) because it became clear that these theories encounter serious difficulties without scalar fields. To explain this let us start from electrodynamics and discuss e.g. electron-electron scattering illustrated by diagram presented in figure 4.9 a. Other processes contributing to this scattering are also possible, for example those described by diagrams 4.9 b-d. Each additional emission and absorbtion of a photon results in a small factor $1/137$ and so the contribution of diagrams b-d should be small. The calculations show, however, that each of these three diagrams gives an infinite contribution into the scattering amplitude because the integrals which appear in the calculations are divergent. Slightly changing the famous Feynman's words one can say that "the corrections are small but infinite".

Higher orders of perturbation theory had been a difficult problem for a long time until the renormalization procedure was found. It was proved that all the infinities can be hidden in electron's mass and charge. These two parameters are to be determined from experiment. The photon mass is zero due to gauge invariance. In other words, all divergent integrals give a contribution either to the electron's mass or charge independently from the type of the phys-

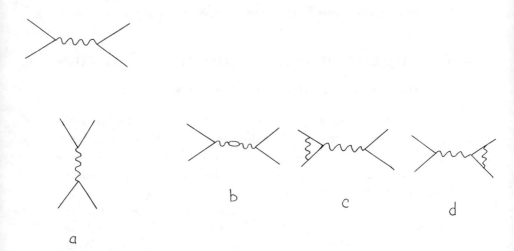

Figure 4.9: Radiative corrections to elastic ee-scattering.

ical process and the structure of diagram. Thus putting together
all the contributions to the electron mass and adding them to the
so-called bare mass of electron m_0, one obtains physical mass of
electron:

$$m_{phys} = m_0 + \alpha \delta m_1 + \alpha^2 \delta m_2 ...$$

where m_0 and δm_i are formally infinite. The sum is, however, finite
and by assumption is equal to the experimentally measured value.

It is clear that this procedure does not look satisfactory but no
alternative have been found [4] on one hand and the results obtained
with the help of this procedure agree very well with experiments on
the other. For example the radiative corrections to the magnetic
moment of the electron can be found .They are described by the
diagrams of fig. 4.10 which modify electron interactions with an
external magnetic field H. The result is:

---

[4]Supersymmetric theories considered in the last years may not have this shortcoming. Infinities in these theories may be cancelled out because of mutual cancellation of bosonic and fermionic contributions to the integrals.

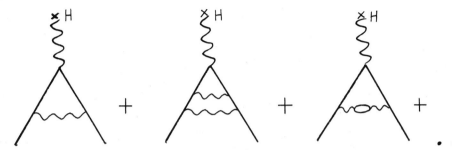

Figure 4.10: Radiative corrections to magnetic moment of electron. The source of external magnetic field is represented by the cross. Waved lines represent virtual photons.

$$\mu = \mu_D \left(1 + \frac{\alpha}{2\pi} - 0.328 \frac{\alpha^2}{\pi^2} + ... \right)$$

This agrees with experiment with accuracy better than 10^{-8}. Not many physical theories demonstrate such an impressive agreement.

The splitting of $2S_{1/2}$ and $2P_{1/2}$ levels in the hydrogen atom is another example of the same kind. These states would be degenerated if there were no quantum corrections . Because of the latter the field is not strictly a Coulomb one, and the photon-electron interaction is modified (figure 4.11). As a result the S-state is predicted to be higher than the P-state by about 1057 MHz. This prediction is in excellent agreement with experiment.

The lesson we learned from electrodynamics is that higher order perturbative corrections (so-called radiative corrections) exist and lead to observational consequences. Radiative corrections to weak interactions are not so important because they are smaller than in electrodynamics and the experimental accuracy is not so high. Still it is desirable to have the theory in which they can be calculated. The above mentioned, local four- fermion theory encounters here a considerable difficulty. This is a so-called non-renormalizable theory. In this theory the higher the order of perturbation, the more infinities arise, and more renormalization constants are needed, not

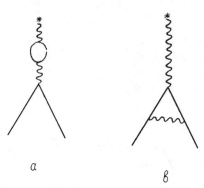

Figure 4.11: Diagrams responsible for the splitting of S and P states in hydrogen atom.

just mass and charge. The reason for this is that local four fermion amplitudes grow with the energy of colliding particles. Thus when higher orders of perturbation theory are calculated and one has to integrate over momenta and energies of virtual particles, the integrals become more and more divergent the higher the order.

When an intermediate boson is responsible for the interaction, then the four-fermion amplitude no longer grows with energy (this is where renormalizability of electrodynamics comes from). For renormalizability the vanishing of the photon mass is also essential. The point is that the amplitude for massive boson emission contains the factor $(E/m)^2$ and still increases with energy. There are no such terms for massless particles because of gauge invariance. The weak interaction intermediate bosons should, however, be massive and therefore the theory is not renormalizable again. The problem is solved by a scalar field.

4.8   Spontaneous   symmetry   breaking and scalar (Higgs) field

The idea of the solution is the following. We shall start from the gauge theory of massless vector fields γ, Z^0, $W^\pm$ and then introduce mass to $W^\pm$ and Z^0 by a dynamical mechanism that changes properties of the vacuum. This is a so-called soft symmetry breaking.

The theory will remain renormalizable since vector fields do not feel vacuum properties when their energies are high. Renormalizability is determined by the behaviour of amplitudes at high energies.

Let us explain the mechanism in more detail. We recall first that the phase of a wave function in quantum mechanics is not observable. Nothing then can depend on the choice of the phase. Formally the Lagrangian ought to be invariant with respect to the transformations:

$$\phi \to e^{i\theta}\phi \tag{4.8}$$

If further a natural assumption is made that phase transformations are made independently in different spacetime points then θ is a function of coordinates. The derivative terms in the Lagrangian, however, break the invariance with respect to transformations (4.8). To restore it a vector field A_μ transforming according to (3.10) must be introduced and covariant derivatives D_μ (3.14) must be substituted for usual derivatives ∂_μ. In fact we have just formulated the principle of gauge invariance. It is successfully used in electrodynamics and also forms a basis of more complicated theories like the theory of electroweak interactions. This procedure for the case of spinor fields, for which the derivative term in the Lagrangian is $\overline{\Psi}\gamma_\mu\partial^\mu\Psi$, defines a fermion current $J_\mu A^\mu = e\overline{\Psi}\gamma^\mu\Psi A_\mu$ and thus the fermion interaction with a vector field. The current J_μ is conserved because of gauge invariance.

The Lagrangian of a scalar field is bilinear in derivatives $(\partial_\mu\phi)(\partial^\mu\phi)$. This term after the substitution $\partial_\mu \to D_\mu$ transforms to:

$$\left|(\partial_\mu - ieW_\mu)\phi\right|^2$$

One of the terms in this expression is:

$$e^2W_\mu^2\phi^2$$

If $\phi \neq 0$ then the factor $e^2\phi^2$ is equivalent to the mass of the W_μ field. In particular it looks exactly like a mass term in the equation

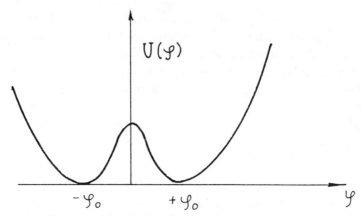

Figure 4.12: The potential of field ϕ that leads to spontaneous symmetry breaking. The point $\phi = 0$ corresponds to unstable equilibrium and the points $\phi = \pm\phi_0$ to the stable equilibrium. The symmetry is broken when $\phi = \pm\phi_0$.

of motion for W-bosons [5]:

$$(\partial^2 + e^2\phi^2)W_\mu = 0 \qquad (4.9)$$

Thus, W boson acquires mass due to interaction with ϕ. The theory proves to be renormalizable if mass is introduced in this way.

This resembles the appearance of a photon mass in plasma. The photon mass in vacuum is zero. Long range electromagnetic interactions are screened in plasma. That means that the photon becomes massive. In the theory we discuss the vacuum is such a medium for W and Z bosons.

We did not yet explain how the field ϕ acquires a constant non-zero value in vacuum, which is the lowest energy state. Let us assume that the potential energy of ϕ is of the form shown in figure 4.12. This potential is what is needed. The lowest energy state is $\phi = \pm \phi_0 \neq 0$. There are two vacuum states so the vacuum is degenerate. The point $\phi = 0$ at which the potential has its local maximum is also an equilibrium point of ϕ. The equilibrium at this point, however, contrary to $\phi = \pm \phi_0$, is not stable. Small

---

[5]The terms describing interaction of W with other fields and in particular the selfinteraction of W (i.e nonlinear terms) are omitted here.

fluctuations of ϕ around zero make ϕ "roll down" to $\phi = \pm \phi_0$. The state $\phi = 0$ is called the false vacuum. The field ϕ is called the Higgs field because it was first introduced to elementary particle physics by Higgs (and also by Englert and Braut and Guralnik, Hagen and Kibble, all in 1964).

Potential energy like that presented in fig. 4.12 has already been long known in ferromagnetic theory. It describes interactions between magnetic moments of atoms. The point $\phi = 0$ corresponds to the state in which all the magnetic moments are directed randomly so that the mean magnetic moment is zero. The state when all magnetic moments are parallel is more favorable from an energetic point of view. It corresponds to non-zero mean magnetic moment or $\phi = \phi_0$ in our picture. The direction of the mean magnetic moment is arbitrary so the vacuum in this model is infinitely degenerate.

Such theories are called theories with spontaneously broken symmetries. The point $\phi = 0$ corresponds to a symmetric state. It is rotational symmetry in the ferromagnetic case. In the scalar field case it was symmetry of the properties of W, Z, and γ. Let us note that masses of all four bosons are zero if symmetry is not broken i.e. $\phi = 0$.

When $\phi \neq 0$ the symmetry is broken. The system goes to this state spontaneously. Vector bosons acquire different masses. Despite the fact that the Lagrangian of the scalar field is invariant with respect to some transformations in internal isospin space the ground state is not. It is analogous to appearance of a special direction in the ferromagnetic case, that is the direction of the magnetic moment and so the rotational symmetry is broken.

Let us note for future use that heating of a ferromagnet destroys the correlation of magnetic moments and restores the symmetry.

The introduction of a scalar field with the peculiar potential of fig. 4.12 is the result of the requirement of renormalizability of massive vector boson theory. In the simplest case the potential can be written as:

$$V(\phi) = \frac{\lambda}{4}(\phi^2 - \phi_0^2)^2 \qquad (4.10)$$

The Lagrangian of field ϕ:

$$L = \frac{1}{2}\left(\frac{\partial \phi}{\partial x_\mu}\right)^2 - V(\phi)$$

describes particles with mass $m = \sqrt{2\lambda}\,\phi_0$ [6].

These particles are called Higgs bosons H or χ of electroweak interaction theory. The exchange of Higgs bosons exactly cancels out the rise of amplitudes with energy due to the nonvanishing masses of the $W^\pm$ and Z^0 and so it ensures renormalizability of the theory.

H bosons have not yet been found experimentally and their mass is unknown. The experimental proof of the existence of the Higgs boson is one of the main tasks of contemporary particle physics.

Complex Higgs fields may form so-called cosmic strings, which are very interesting objects from the cosmological point of view. After spontaneous symmetry breaking in the theory with potential $V(\phi) = \lambda(|\phi|^2 - \phi_0^2)^2$ the phase of ϕ is not correlated in different causally disconnected regions and it may happen that the phase variation along a closed path is non-zero $\Delta\phi = 2\pi n$. This means that a topologically stable object, a string or a vortex line is formed. To be more exact if $n > 1$ there are n strings formed. In the centre of a string $\langle\phi\rangle = 0$ and far from it $|\langle\phi\rangle| = \phi_0$. These objects have microscopical transverse size $L = (\sqrt{\lambda}\phi_0)^{-1}$ and macroscopical length. They may play an important role in formation of the large scale structure of the Universe serving as seeds for rising density fluctuations.

---

[6]This subject is discussed in more detail in sec.1 of chapter 6.

Chapter 5

Cosmological constant

The de Sitter metric of exponentially expanding space plays an important role in the theory of the early Universe and in solutions of the flatness and horizon problems. This metric is a homogeneous and isotropic solution of the General Relativity equations in vacuum with the so-called cosmological constant. The cosmological constant was first introduced to the gravity equations by Einstein in order to obtain a stationary cosmological model in which the attraction of matter is compensated by a Λ-term. Einstein rejected the idea as soon as it appeared that the Universe is expanding. We believe now, however, that the expansion of the Universe at early stages was determined by the Λ-term or by an equivalent state of matter.

The observations show that the contemporary value of cosmological constant is either very small or equal to zero. On the other hand quantum field theory as well as analysis of phase transitions in the early Universe predict a very large value of it. This tremendous contradiction between the predictions and observations presents the serious problem of cosmological constant in the Universe.

5.1 Definition and physical meaning

An extra term, which does not violate general covariance and covariant energy-momentum conservation, $D_\mu T^\mu_\nu = 0$, can be introduced to the Einstein equations (3.21):

$$R_{\mu\nu} - 1/2Rg_{\mu\nu} + \Lambda g_{\mu\nu} = 8\pi G T_{\mu\nu} \qquad (5.1)$$

Λ is called the cosmological constant. In contrast to the first two terms in the left-hand side of this equation $\Lambda g_{\mu\nu}$ does not vanish when $g_{\mu\nu}$ is the metric of flat spacetime. It can be regarded as a contribution to the energy-momentum tensor. It is covariantly conserved because:

$$D_\alpha g_{\mu\nu} = 0 \qquad (5.2)$$

and has the same value in all inertial coordinate frames. Since the vacuum is invariant with respect to Lorentz transformations $\Lambda/8\pi G$ is sometimes called the vacuum energy density. Quantum field theory gives deep physical reasons for this. The ground state (vacuum) energy in quantum theory is generally nonvanishing. The vacuum expectation value of the energy-momentum tensor in flat space time is proportional to $g_{\mu\nu}$:

$$T_{\mu\nu}^{vac} = \langle vac|T_{\mu\nu}|vac\rangle = \rho_{vac}g_{\mu\nu} \qquad (5.3)$$

This will be discussed in more detail in section 4.

Let us note that the energy-momentum tensor of an ideal fluid is:

$$T_{\mu\nu} = (p + \rho)u_\mu u_\nu - pg_{\mu\nu} \qquad (5.4)$$

It coincides with vacuum tensor (5.3) when $p = -\rho$ i.e. the above mentioned relation $p = -\rho$ is a property of the vacuum tensor (5.3). We will prove that the relation $p = -\rho$ is invariant with respect to Lorentz transformations. Although the result is quite obvious these simple calculations will be presented because of their special importance.

A second rank tensor and in particular the energy momentum tensor, is transformed according to the formulae:

$$T_{00} = (1 - v^2)^{-1}(T_{00}' - 2vT_{01}' + v^2T_{11}')$$

$$T_{11} = (1 - v^2)^{-1}(T_{11}' - 2vT_{01}' + v^2T_{00}')$$

when the primed coordinate system (K') is moving along axis OX of the nonprimed system (K). The other components can be divided into two groups: the components T_{22}, T_{23}, T_{33} are not transformed i.e. $T'_{22} = T_{22}$ etc., components T_{12}, T_{13} and T_{01}, T_{02}, T_{03} are transformed as:

$$T_{12} = (1 - v^2)^{-1/2}(T'_{12} - vT'_{02})$$

$$T_{02} = (1 - v^2)^{-1/2}(T'_{02} - vT'_{12})$$

$$T_{01} = (1 - v^2)^{-1/2}\big((1 + v^2)T'_{01} - v(T'_{00} + T'_{11})\big)$$

and analogously for T_{03} and T_{13}.

Let $T'_{\mu\nu}$ in the primed system be of the form (5.3). Since $T'_{0i} = (\epsilon + p)v'_i = 0$, there is no energy flux in (K'). Nondiagonal stresses T_{12}, T_{13}, T_{23} are also zero because pressure is isotropic. The only nontrivial transformations are:

$$T_{00} = (1 - v^2)^{-1}(T'_{00} + v^2 T'_{11})$$

$$T_{11} = (1 - v^2)^{-1}(T'_{11} + v^2 T'_{00})$$

$$T_{01} = -v(1 - v^2)^{-1/2}(T'_{00} + T'_{11})$$

After inserting $T'_{00} = \rho$ and $T'_{11} = -\rho$ one obtains $\rho = \rho'$ and $p = -\rho$. The conclusion is that the relation $p = -\rho$ is Lorentz invariant. In fact it follows from the Lorentz invariance of the Minkowski space.

Let us note that the Minkowski solution is not consistent with $\Lambda \neq 0$. Any spacetime that is a solution of General Relativity equations with $\Lambda \neq 0$, must be curved. If there is no ordinary matter, i.e. $T_{\mu\nu} = 0$ then this solution is called the de Sitter solution.

5.2   The de Sitter solution

The general form of the spacetime interval in a homogeneous and isotropic Universe (2.1) is not affected by the addition of the Λ-term. This is evident, since the Λ-term corresponds to a special choice of $T_{\mu\nu}$. The equations governing scale factor evolution are:

$$\frac{1}{2}\left(\frac{da}{dt}\right)^2 - \frac{4\pi G\rho a^2}{3} = \frac{1}{6}\Lambda a^2 - \frac{k}{2}$$

This is an analogue of equation (2.2), or the energy conservation law. The equation expressing acceleration $\ddot{a}/a$ through the total mass is also changed:

$$\ddot{a} = \frac{1}{3}\Lambda a - \frac{4\pi G}{3}(\rho + 3p)a$$

where p and ρ are the pressure and energy density of matter only, while the vacuum quantities are included in the terms proportional to Λ.

Let us find the physical meaning of the Λ term. To this end the following simple model will be used. Let us consider a small space region bounded by the sphere with radius r. The distance between the centre and any point on its surface is determined by the equation:

$$\frac{d^2(ar_0)}{dt^2} = \frac{1}{3}\Lambda ar_0 - \frac{GM}{(ar_0)^2} \tag{5.5}$$

Here M is the total mass of matter inside the sphere which is equal to $(4\pi/3)(ar_0)^3(\rho + 3p)$. It is clear now that two forces act on a particle lying on the surface of the sphere. The first is the ordinary Newton attraction:

$$F_N = -GM/(ar_0)^2$$

and the second, caused by Λ-term, is also a gravitational force but rather unusual:

$$F_\Lambda = \frac{1}{3}\Lambda ar_0$$

The field of such forces is global and homogeneous. The value of the force depends on the distance between interacting particles as (ar_0). The further away are the particles, the larger is the force. It formally resembles the interaction between quarks discussed in chapter 4.

At the present time the Λ-term is either very small or equal to zero so the forces proportional to r are not essential in ordinary astronomy. They can however appear on scales comparable with the horizon size. That may be important in cosmology and that is why Λ is called cosmological constant.

When Λ is positive the F_Λ force is a repulsive force proportional to the distance between particles. When Λ is negative then additional gravitational attraction arises. This is the physical reason why the scale factor in the de Sitter space corresponding to closed, flat or open cross sections behaves differently from that in the Friedman space. In particular, even in the open de Sitter space the scale factor may not increase infinitely but may reach a maximum and then fall down to zero.

The variation of energy density with time is determined by the dependance $p = p(\rho)$ and in the three most important cases is given by the equations:

$$\frac{d\rho}{dt} = -3H\rho, \quad \text{if} \quad p = 0$$

$$\frac{d\rho}{dt} = -4H\rho, \quad \text{if} \quad p = \rho/3$$

$$\frac{d\rho}{dt} = 0, \quad \text{if} \quad p = -\rho$$

The fact that ρ_{vac} does not decrease, can change the predictions about the future of the Universe.

When Λ is negative and there is an additional attractive force, the Universe will necessarily contract after some expansion period. This is obvious, since the cosmological attraction force increases as the scale factor: $F_\Lambda \sim \Lambda a$. If Λ is positive and there is the cosmological repulsion force the choice is richer than in the case of negative Λ. The notion of open, flat, or closed Universe becomes ambiguous. In this case the four dimensional spacetime manifold can be split into three dimensional space and time in different ways so that the three dimensional space can be either closed, flat or open with the same set of the cosmological parameters. This surprising property of the de Sitter space will be discussed later.

The topological difference between the de Sitter and Friedman spaces is, that when $\Lambda = 0$ the three dimensional closed world $t = const$ is closed also in time. If $\Lambda \neq 0$ it is open in time, and as we shall see in what follows another space cross- section corresponding to another constant time coordinate $t' = const$ can be open (in space).

Let us discuss the gravity equations with a Λ-term. We first consider the simple case of $p = 0$ and of course $\Lambda = 0$. The equations determining scale factor evolution can be written as follows:

$$\ddot{a} = -\frac{GM_1}{a^2}$$

$$\frac{\dot{a}^2}{2} = \frac{GM_2}{a} - \frac{k}{2}$$

When $p = 0$ the same mass $M_1 = M_2 = V\rho$ enters both equations. The first equation is the Newton equation of motion, and the second one is the energy conservation law. The first can be obtained by differentiating the second and assuming that M=const.

If $p \neq 0$ masses M_1 and M_2 become different, $M_1 = V(\rho + 3p)$ and $M_2 = V\rho$. The equations now are in concordance with the well known law of energy variation $dM_2 = -pdV$.

In the particular case $p = -\rho$ the equation of motion is:

$$\ddot{a} = \frac{8\pi G\rho}{3}a \qquad (5.6)$$

It has the following solutions:

$$\exp\left(\pm\sqrt{\frac{8\pi G\rho}{3}}\, t\right)$$

The concrete form of the solution depends on k which determines the difference between kinetic and potential energies.

If Λ is positive and $k = 0$ there is the pure exponential solution:

$$a(t) = a_0 \exp\left(\sqrt{\frac{\Lambda}{3}}\, t\right) \tag{5.7}$$

The Hubble parameter in such a space is constant (this is the case when the Hubble constant is really constant unlike in the Friedman space where it depends on time):

$$H = \frac{\dot{a}}{a} = \sqrt{\frac{\Lambda}{3}} \tag{5.8}$$

The redshift in the de Sitter space with $k = 0$ is:

$$z = \exp\left(\sqrt{\frac{\Lambda}{3}}\,(t_0 - t)\right) - 1$$

In the case of $k = \pm 1$ the scale factor depends on time as hyperbolic sine or cosine. When $k = +1$, which in the case of the Friedman universe would correspond to closed spacetime with topological properties of a sphere, the scale factor is:

$$a(t) = a_0 \cosh\left(\sqrt{\frac{\Lambda}{3}}\, t\right) \tag{5.9}$$

Here a_0 is not an arbitrary constant. It is expressed in terms of Λ:

$$a_0 = \sqrt{\frac{3}{\Lambda}}$$

In the case of $k = -1$ the solution of equation (5.6) becomes:

$$a(t) = \sqrt{\frac{3}{\Lambda}} \sinh\left(\sqrt{\frac{\Lambda}{3}}\, t\right) \tag{5.10}$$

The Hubble parameter in the de Sitter space with $k \neq 0$ is not constant. It is equal to:

$$H = \sqrt{\frac{\Lambda}{3}} \tanh\left(\sqrt{\frac{\Lambda}{3}}\, t\right) \quad \text{when} \quad k = +1$$

$$(5.11)$$

$$H = \sqrt{\frac{\Lambda}{3}} \coth\left(\sqrt{\frac{\Lambda}{3}}\, t\right) \quad \text{when} \quad k = -1$$

The difference between the three solutions presented above quickly becomes negligible when $\sqrt{\Lambda/3}\, t \gg 1$.

The de Sitter space has a very interesting property: it is symmetric to the same high degree as flat spacetime and hence the time axis can be chosen to a very large extent arbitrarily. There is no such freedom in the Friedman Universe. Four dimensional spacetime can be split into three dimensional space and a time axis in many possible ways. The physical criterion to determine a three dimensional hypersurface in the homogeneous Universe is that the energy density be constant at a fixed time. The energy density in the Friedman Universe changes with time and so the time axis can be defined as the axis which is perpendicular to the hypersurface $\rho = const$. This criterion is not valid in de Sitter space since $\rho = const$ not only in space but also in time. All three solutions with k=+1, 0, and -1 can be obtained by choosing different directions of time axis in this four dimensional manifold. The closed solution (k=+1) covers all the four dimensional manifold, while the flat and open ones are only parts of it. This property will be illustrated, in what follows, by the example of a two dimensional surface in three dimensional pseudoeuclidean space.

The invariance group and topology of the de Sitter space can be described in the following way. Let us imagine the de Sitter space as a hyperboloid in the five dimensional space with the interval:

$$ds^2 = dt^2 - dx^2 - dy^2 - dz^2 - dv^2$$

where t, x, y, z are ordinary four dimensional coordinates and v is a new one which allows to consider a four dimensional curved manifold embedded in a five dimensional Minkowski space.

The equation of a hyperboloid in 5-dimensional space is:

$$t^2 - x^2 - y^2 - z^2 - v^2 = -r_0^2 \qquad (5.12)$$

Introducing five dimensional spherical coordinates we can parametrize this equation by new variable τ which will later play the role of time:

$$t = r_0 \sinh(H\tau), \quad x = r_0 \cosh(H\tau)\sin(r)\sin(\theta)\cos(\phi)$$

$$y = r_0 \cosh(H\tau)\sin(r)\sin(\theta)\sin(\phi)$$

$$z = r_0 \cosh(H\tau)\sin(r)\cos(\theta), \quad v = r_0 \cosh(H\tau)\cos(r)$$

Functions $(t, x, y, z,$ and $v)$ describe a hyperboloid or in other words satisfy equation (5.12). The interval on the hyperboloid expressed in terms of four coordinates (τ, r, θ, ϕ) is:

$$ds^2 = d\tau^2 - H^{-2}\cosh^2 H\tau\left(dr^2 + \sin^2 r(d\theta^2 + \sin^2\theta d\phi^2)\right)$$

where $r_0^2 = H^{-2}$ i.e. it is the de Sitter solution with k=+1. This is a complete coordinate system that covers all the de Sitter manifold.

It is not easy to imagine a four dimensional hyperboloid embedded in five dimensional space so a two dimensional hyperboloid in three dimensional pseudoeuclidean space will be discussed in what follows.

The interval in three dimensional pseudoeuclidean space is chosen as:

$$ds^2 = dt^2 - dx^2 - dv^2$$

where t is a "three dimensional time" and v is an additional coordinate. A hyperboloid in this space is described by the equation analogous to eq. (5.12):

$$t^2 - x^2 - v^2 = -r_0^2 \tag{5.13}$$

In the same way as in five dimensions, spherical coordinates r and a parameter τ are introduced:

$$t = r_0 \sinh(\tau/\tau_0); \quad x = r_0 \cosh(\tau/\tau_0) \sin r; \quad v = r_0 \cosh(\tau/\tau_0) \cos r$$

These coordinates automatically satisfy equation (5.13). The two dimensional interval on the surface of the hyperboloid is:

$$ds^2 = d\tau^2 - r_0^2 \cosh^2(\tau/\tau_0) dr^2$$

This interval resembles the metric of a closed space. Although the coordinate r formally changes from $-\infty$ to ∞ it is cyclic with period 2π.

Lines obtained by crossections of the hyperboloid with planes t=const describe one dimensional space in the model universe we are considering. These lines are circles with length $l = 2\pi r_0 \cosh(\tau/\tau_0)$ (these are one dimensional models of closed space). The shortest circle or equivalently, the smallest volume of Universe is obtained at $\tau = 0$ or on the crossection $t = 0$. The smallest volume crossection is not invariant with respect to Lorentz transformations. Transition to the system moving relative to the initial one with speed β gives:

$$t = \frac{t' - \beta x'}{\sqrt{1 - \beta^2}}, \quad x = \frac{x' - \beta t'}{\sqrt{1 - \beta^2}}, \quad v = v'.$$

The hyperboloid equation in new coordinates is of the same form:

$$t'^2 - x'^2 - v'^2 = -r_0^2$$

and the interval also remains the same:

$$ds^2 = d\tau'^2 - r_0^2 \cosh^2(\tau'/\tau_0) dr'^2$$

The circle of minimum length $l = 2\pi r_0$ in new coordinates is also defined by the condition $\tau' = 0$ but this circle is not the same as the former one. The plane $t' = const$ satisfies the equation $x + \beta t = const$ in the former coordinates (t,x). The equation of planes containing the minimal radius circle is:

$$t + \beta x = 0$$

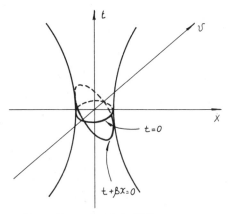

Figure 5.1: Sections of two dimensional de Sitter space corresponding to a spacially closed world. Each circle on the hyperboloid surface with centre at point 0 has the same length.

Here β is the angle between two circles of minimum length in these two coordinate systems as shown in figure 5.1. Infinitely many circles with minimum length exist. This is connected with the non-definitness of the metric sign. This phenomenon is well known in Minkowski space. The line $x = const$ corresponds to the maximum Δt. It is instructive in this connection to recall the famous twins paradox: those who move are younger. In this simple model of de Sitter space we shall show how flat, open, and closed solutions can be obtained from a single hyperboloid.

So far we have considered the closed de Sitter universe, and now the open space will be obtained. To this end another parametrization of the hyperboloid and also another definition of time is needed. Let us consider a section of the hyperboloid by a plane $v = const$ perpendicular to Ov axis. The sections are hyperbolas described by $t^2 - x^2 = v^2 - r_0^2$. The hyperboloid can be parametrized as follows:

$$t = r_0 \sinh(\tau/\tau_0) \cosh(\chi)$$
$$x = r_0 \sinh(\tau/\tau_0) \sinh(\chi); \; v = r_0 \cosh(\tau/\tau_0)$$

and the metric on the hyperboloid is:

$$ds^2 = d\tau^2 - r_0^2 \sinh^2(\tau/\tau_0) d\chi^2$$

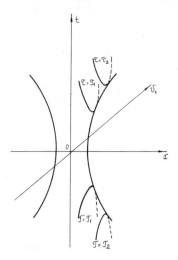

Figure 5.2: Sections of two dimensional de Sitter space corresponding to open spacially curved world.

This is an open Universe metric.

The space section of this world is the hyperbola formed by the intersection of the hyperboloid with the plane perpendicular to the axis Ov. The length of the hyperbola is infinite and so the world is open. Time is counted along the Ov axis as shown in figure 5.2. Along these hyperbolas $\tau = const$ so each of them is a spacelike hypersurface at different times. Let us stress once more that the choice of time axis is conventional.

Now we discuss how flat space can be obtained but do not present an explicit parametrization. A hyperboloid is known to be formed by straight lines as shown in the figure 5.3. Such a line corresponds to a flat spacelike section in the de Sitter space. The time is counted perpendicularly to these straight lines. These considerations permits us to write down explicitly the metric of the de Sitter world with flat space sections.

Thus one can get different de Sitter worlds with open, closed, or flat spaces by choosing different spacetime crossections. Let us recall that the closed one covers all the hyperboloid, while the other two do not. The latter two are said to be geodesically not complete.

The possibility of transforming, say, the flat de Sitter world into

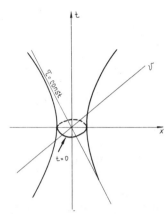

Figure 5.3: Sections of two dimensional de Sitter space corresponding to open spacially flat world.

a closed one is connected with the symmetry group of de Sitter space, or the group of movements. This group, like the one in Minkowski space, is a 10 parameter group of transformations. It means that there are 10 fundamental transformations which generate any transformation in this group. There are three space rotations since the space is isotropic. There are also three translations along three space coordinate axes, which reflects the homogeneity of de Sitter space. There are, four more transformations apart from the six already mentioned. The first three of them are complex spacetime rotations. From the physical point of view these correspond to transition to a coordinate system moving with a constant velocity. The last one is the translation along the time axis. These ten transformations generate the symmetry group of de Sitter space. It is the ten parameter group of Lorentz transformations in five dimensional Minkowski space. This group is called SO(4,1).

Let us discuss now the kinematics of de Sitter space. Namely we compare the horizons in this and Friedman spaces. The kinematical properties can help to understand the flatness problem and how the causally connected regions evolve.

The definition of the horizon was introduced in section 2 of chapter 2. The horizon is the distance a light ray travels from time $t_e = 0$ when it was emitted till the time t_0 when it was observed:

$$l_h = a(t) \int_{t_e}^{t_0} \frac{dt'}{a(t')}$$

The scale factor in the spatially flat Universe is:

$$a(t) = a_0 \exp(Ht)$$

and the horizon depends on t_0 as:

$$l_h = H^{-1} \exp(Ht_0) \left(1 - \exp(-Ht_0)\right) \quad \text{if} \quad t_e = 0.$$

Although l_h grows very quickly, the scale factor grows even faster. The distance to the horizon in r-coordinates, $r = l/a(t)$, approaches a constant value, namely:

$$r_h \rightarrow (a_0 H)^{-1} \tag{5.14}$$

Only points that were initially inside a sphere with radius H^{-1} can ever exchange signals. If, in particular, $H = t_{Pl}^{-1}$ at the moment of creation of the Universe then only points inside spheres with radii t_{Pl} can be causally connected. In r-coordinates this connection will always be inside the sphere with $r < t_{Pl}$. Points initially further apart than $1/H$ will never exchange any signals. Such points at present are exponentially far from each other. For $Ht_0 > 100$ the causally connected region is much larger than the modern horizon size.

Let us compare our results with the behaviour of the horizon in the Friedman Universe. For the latter the distance to horizon depends on the time of observation as:

$$l_h = t_0 const, r_h = l_h/a(t) = \begin{pmatrix} t^{1/2} & \text{in} & \text{the} & \text{RD} & \text{case} \\ t^{2/3} & \text{in} & \text{the} & \text{MD} & \text{case} \end{pmatrix}$$

In the infinite future there will be the causal connection even between the points which were infinitely far away (in r coordinates). This is the principal difference between Friedman and de Sitter Universes. No matter how strange it sounds a de Sitter stage in the early Universe explains the causal connection in the presently observed Friedman Universe.

Let us discuss the evolution of causally connected regions in the de Sitter space with k=0. The coordinate r is the so called Lagrange coordinate of a particle[1]. Roughly speaking, different values of r correspond to different particles. The distance between particles is, of course, constant in r coordinates but the physical distance, measured, say, by a light ray, increases.

Let us now consider a numerical example of the evolution of causally connected regions in de Sitter space. It is assumed that a de Sitter Universe was born at $t = t_{Pl}$ with Hubble parameter $H = t_{Pl}^{-1}$ and causally connected regions of size $l = l_{Pl}$. These regions grow $e^{70} = 10^{30}$ times and reach 10^{-3}cm in 70 Hubble times. This is exactly the size needed to explain causal connection of our Universe. However, $Ht = 70$ is by no means a limit to inflation. The existing models naturally give $Ht \gg 70$ and thus the size of a region that evolved from $l = l_{Pl}$ is much greater than the present horizon.

The size of the horizon in the de Sitter space with closed spatial sections, differs from (5.14) only by a numerical factor. If the scale factor $a(t) = a_0 \cosh(Ht)$ then the horizon size in r coordinates is:

$$r = \pi/(2Ha_0) \qquad (5.15)$$

when $Ht_0 \gg 1$. It is quite different from the case of open de Sitter space. For the latter the scale factor is:

$$a(t) = a_0 \sinh(Ht)$$

and the integral defining horizon size is logarithmically divergent in the lower limit:

$$a_0 r_h = \int_{t_e}^{t_0} \frac{dt}{\sinh(Ht)} =$$

$$= H^{-1} \left(\ln \frac{e^{Ht_0} - 1}{e^{Ht_0} + 1} + \ln \frac{e^{Ht_e} + 1}{e^{Ht_e} - 1} \right) \approx$$

$$\approx H^{-1} \ln(2Ht_e) \qquad \text{for} \qquad t_e \to 0$$

---

[1]By definition the Lagrange coordinates are formed by trajectories of test particles.

Formally one might conclude that the whole space is causally connected, since the horizon size grows infinitely as $t_e \to 0$. This result, however, is nonphysical. The solution to this paradox is that the de Sitter stage in the real world begins at t_{Pl} or later. So instead of zero one should substitute the time of the beginning of de Sitter stage t_e.

Knowing the kinematical properties of the de Sitter Universe, we turn now to the flatness problem. To this end we consider the behaviour of kinetic and potential energies of the Universe. In an empty Universe with positive cosmological constant there are only repulsive forces :

$$F_\Lambda = H^2 a$$

Let us recall that a has the dimension of length. Thus potential energy per unit mass in such a Universe is:

$$U = H^2 a^2 / 2$$

and the kinetic energy per unit mass is:

$$E = (da/dt)^2 / 2$$

Both E and U grow exponentially in the de Sitter Universe:

$$E = \cosh^2 Ht; \quad U = \sinh^2 Ht$$

but their difference remains constant:

$$E - U = k/2$$

After a sufficiently long time both energies become much larger than their difference and are approximately equal to each other:

$$E \approx U \gg |k/2|$$

The solution of the flatness problem is based on these considerations.

The parameter Ω is defined as the ratio of the matter density in the Universe to the critical matter density. It can equivalently be written as the ratio of potential to kinetic energy. Now it is clear that $(1 - \Omega)$ is a ratio of a constant difference (E-U) to E which grows very quickly. Therefore, regardless of what was the initial value of $(\Omega - 1)$ in the beginning of the de Sitter stage, it approaches zero if this stage lasts long enough. As we have already mentioned the duration of the de Sitter stage t_0 is to satisfy the condition $Ht_0 > 70$ so that the cosmological problems discussed above can be solved.

The value of Ω at the moment when Friedman expansion begins must be close to unity, with the accuracy of about 10^{-60}. Thus the problem of flatness of the early Universe is solved.

The evolution of ordinary matter with equation of state $p = 0$ or $p = \rho/3$ in de Sitter space can be easily understood. Energy density decreases as a^{-3Ht} in the first case and as a^{-4Ht} in the second. As the scale factor grows 10^{30} times the density decreases 10^{120} or 10^{90} times respectively. In other words the ordinary matter density in de Sitter space quickly decreases and becomes negligible. Therefore, the expanding Universe with $a \sim \exp(Ht)$ can be discussed without ordinary matter when $t > H^{-1}$.

The de Sitter solutions with k=1 and k=0 have one more important property. They are not singular i.e. the scale factor a(t) is never equal to zero[2]. This is an important difference from the Friedman's solution where $a(t) \to 0$ when $t \to 0$. This point is called the initial singularity, there $\rho \to \infty$ and $T \to \infty$. The real cosmology however is not the de Sitter one. The exact de Sitter solution with $\Lambda \neq 0$ and $T_{\mu\nu}$ is stable and can not go into the Friedman one. Since the Universe expands according to the Friedman law starting from $t = 1s$ at least, the de Sitter solution must be approximate. This is indeed realized in the inflationary Universe models.The vacuum energy of the scalar field transforms into energy of particles and the expansion regime changes into the Friedman one.

On the other hand, if Λ is small and nonzero then after a sufficiently long time the Friedman expansion can change to the de

---

[2]The theorems proving that cosmological singularity is unavoidable are based on the energodominance condition $\rho + 3p > 0$, which in the case considered is evidently violated.

Sitter one. This will necessarily take place if the Universe is open. Note that, even though the Universe would expand exponentially, small bound systems, like galaxies and their clusters, would not, since gravitational attraction on small scales is greater than anti-gravitation due to Λ term.

Turning back to the problem of the initial singularity we would like to note that the principle of rejecting singular solutions in physics is not to be respected. Singular solutions are very often an adequate approximation to reality. Shock waves, which are singular solutions in ideal fluid hydrodynamics present a good example. Probably in cosmology also we must look for the physical mechanisms of deviation from the ideal case.

5.3   Comments on the relation $p = -\rho$

Let us discuss what properties a medium would possess if the equation of state $p = -\rho$ were valid. Let this equation be written in parametric form:

$$p = p(n); \quad \rho = \rho(n)$$

where $n \sim 1/V$ is the particle number density in the medium. Using the well known thermodynamical relation:

$$dE = -pdV \tag{5.16}$$

and taking into account that $E = \rho/n$, one easily obtains that:

$$-(d\rho/dn)\not{x} = -(\rho + p)/n$$

This implies, in particular, that $\rho \sim n^{4/3}$ when $p = \rho/3$ and $\rho \sim n$ when p=0.

The case $p = -\rho$ is evidently degenerate. In this case, neither p nor ρ depend on n. There is no way to change the value of energy density because there are no parameters on which it depends. In place of a continuum of values of p and ρ connected by the relation $p = p(\rho)$ there is only one point $p = -\rho = const$.

In an infinite medium of this kind there are no pressure gradients. If it were not for gravity it would be stationary . If the system is finite, a shell compensating the pressure is needed to stabilize it.

The gravitational interaction of this medium proves to be repulsive when $\rho > 0$ i.e this medium antigravitates. The gravitational repulsion appears only inside the region with the relation $p = -\rho$. When a piece of matter with negative pressure is situated in an empty space where $p = \rho = 0$ it will interact with external objects just like an ordinary gravitating body. To understand this let us consider a simplified stationary case. External forces must exist to keep such a piece of matter in equilibrium. To calculate the gravitational potential created by this body in an external space one must take into account not only $T_{\mu\nu}$ given by equation (5.3), but also additional terms connected with the forces maintaining the equilibrium. It can be easily proved that volume integrals of purely spatial or mixed components of the energy-momentum tensor of a body vanish in the stationary state. Thus, the gravitational field is determined by the total mass of the system, as one would naively expect:

$$m = \int T_{00} dV > 0$$

Regardless of the equation of state of a body any finite stationary object gravitationally attracts external bodies. The essential point is however that the body is finite and not that it is stationary. In this case spatial and time derivatives $(\partial\phi/\partial t)^2$, $(\partial\phi/\partial x)^2$ give a nonvanishing contribution to the energy-momentum tensor ensuring positive definiteness of its total mass.

As an example let us note that negative pressure can be obtained in solid bodies stretched by external forces. Its value is however not as big as it is considered in cosmology: $p \ll \rho$.

Let us stress once more that the energy density of a system with $p = -\rho$ is constant when volume is changed. It can be also proved using the thermodynamical relation (5.16). By definition $E = V\rho$. Differentiating both sides of this equation one obtains:

$$dE = \rho dV + V d\rho \tag{5.17}$$

Comparing the right-hand sides of equations (5.16) and (5.17) we get:

$$V d\rho = -(\rho + p)dV$$

Now it is clear that variation of energy density with volume $dV \neq 0$ vanishes ($d\rho = 0$) only if $p = -\rho$. Therefore energy density is constant in an expanding Universe with the equation of state $p = -\rho$.

5.4   The problem of Λ - term in the contemporary universe

As we have already mentioned, the discovery of the expansion of the Universe has made unnecessary the introduction of a cosmological constant into the gravity equations. Moreover, astronomical observations indicate that, if nonzero, it is very small. The limit expressed as an equivalent vacuum energy density is:

$$\rho_{vac} < 10^{-29} \text{g/cm}^3 \qquad (5.18)$$

This is not a very strong upper bound from the point of view of cosmology. It is of the same order of magnitude as the critical energy density. But if considered in terms of energy scales characteristic of elementary particle physics this quantity is fantastically small. To illustrate this, it is convenient to rewrite the bound (5.18) in different units:

$$\rho_{vac} < 10^{-47} (\text{GeV})^4 \qquad (5.18')$$

A natural question to ask is what is the relation between the cosmological constant and elementary particle physics. The answer is that the ground state, i.e. vacuum energy in quantum field theory is not zero but has the value of m^4 where m is a characteristic particle physics mass parameter. The value of m depends on the particular model but in any known model $m > 1$ GeV. There might be accidental compensation of different contributions to ρ_{vac} but there is little chance for compensation with an accuracy of one part in 10^{50}.

The appearance of a Λ term resembles the appearance of mass due to quantum corrections i.e. renormalization. Gauge invariance implies that the photon mass is zero both in classical and in quantum physics, at least if the coupling constant is sufficiently small. No principle implying the vanishing of the cosmological constant has yet been found. Quantum field theory shows that anything that is not forbidden is allowed. The cosmological constant seems to be an exception to this rule.

In order to explain how quantum theory leads to non zero vacuum energy let us recall that ground state energy of a quantum mechanical oscillator is not zero but $\omega/2$. (This is connected with Heisenberg's uncertainty principle.) There is an important difference between quantum and classical mechanics since in classical mechanics the minimum energy of a particle in the potential $u = m\omega^2/2$ is zero.

Now as a step towards quantum field theory we turn from one particle quantum mechanics to the quantum solid state physics. A solid body consists of a large number of particles. The energy of their small oscillations is:

$$U = \frac{1}{2} \sum m_j^2 x_j^2 + \frac{1}{2} k \sum (x_j - x_{j+1})^2 \qquad (5.19)$$

where the first term is the electron-ion interaction and the second one describes the interaction between neighboring electrons. The eigenstate problem can be solved by diagonalization of the interaction matrix i.e. by transition to normal coordinates. This is achieved by a Fourier transformation. The ground state energy of a solid is equal to half of the sum of normal mode frequencies. Since the energy of an oscillator is nonvanishing so is the ground state energy of a solid even if the temperature is zero. This energy is called the zero mode energy. It is observed in experiments. For example isotopes Li^6 and Li^7 have the same atomic interaction energy but different zero mode energies. This leads to different temperatures of vaporization. This effect is used in isotope separation.

Quantum field theory resembles very much solid state theory. In classical theory a field is described by one function $\phi(x,t)$ (scalar field) or by four functions A_μ (vector field) etc... which determine the state of the field at every spacetime point. Transition to quantum field theory is made as in the case of one particle quantum mechanics by a change from numerical to operator functions. In this sense quantum field theory may be considered as the quantum mechanics of a system with infinitely many degrees of freedom. For example the Hamiltonian of a free scalar field can be written as:

$$H = 1/2 \int d^3x[\pi^2(x,t) + (\nabla\phi(x,t))^2 + m^2\phi^2] \tag{5.20}$$

which is similar to expression (5.19) for the electron energy in a solid. The transition to normal coordinates in this case also can be done by a Fourier transform:

$$\phi(x,t) = \int \frac{d^3k}{\sqrt{(2\pi)^3 2\,\omega_k}} \,[a(k)e^{ikx-i\omega t} + a^+(k)e^{-ikx+i\omega t}]$$

where $\omega_k = \sqrt{k^2 + m^2}$ while $a(k)$ and $a^+(k)$ are the creation and annihilation operators of a particle with momentum k. In terms of these operators the Hamiltonian is diagonal :

$$H = 1/2 \int d^3k\,\omega_k[a^+(k)\,a(k) + a(k)\,a^+(k)]$$

This is the Hamiltonian of a set of noninteracting oscillators with frequencies ω_k. The ground state of such a system has nonvanishing (infinite) energy density:

$$\epsilon = \frac{1}{2}\frac{4\pi}{(2\pi)^3} \int k^2\omega_k dk$$

It can be visualized as the energy of particles which appear and disappear in vacuum (see diagram in figure 5.4).

The result is more transparent when integration is changed to summation according to the formulae :

$$\int dk = \sum_k \Delta V_k$$

$$\delta^3(k - k') \rightarrow \frac{\delta_{kk'}}{\Delta V_k}$$

$$a_k = \sqrt{\Delta V_k}\; a(k)$$

Figure 5.4: Virtual particles that appear and disappear in vacuum. The vacuum energy is altered by their presence.

The operators a_k and a_k^+ commute according to the relation:

$$[a_k, a_{k'}^+] = \delta'_{kk}$$

With this substitution Hamiltonian (5.20) can be rewritten as:

$$H_b = \sum \omega_k (a_k^+ a_k + 1/2) \qquad (5.21)$$

Remembering that the vacuum is a particle free state i.e. $a_k|vac\rangle = 0$ we get for the vacuum energy $\langle vac|H|vac\rangle = \sum \omega_k/2$. This infinite energy was usually not taken into account since it was considered as not observable and particle energies were counted from this infinitely high level. This point of view is selfconsistent if gravity is not taken into consideration. The latter, however, is sensitive to every form of energy including that of the vacuum. The energy-momentum tensor of the vacuum has the form $T_{\mu\nu} = \rho_{vac}g_{\mu\nu}$ and so vacuum energy is equivalent to cosmological constant.

Equation (5.21) holds for bosons (hence the index b). Turning to fermions we must remember that their creation and annihilation operators anticommute. Hence the sign of the second term in the Hamiltonian must be changed:

$$H_f = \sum \omega_k (a_k^+ a_k - 1/2)$$

This change of sign leads to a downward displacement of the fermion oscillatory potential by ω_k relative to the potential of bosons with the same mass (these potentials and their ground states are schematically shown in figure 5.5)[3]. We believe now that there

---

[3]Gravitational effects connected with the Dirac sea was noted by G.Gamow in his letter to A.F.Joffe in 1930.

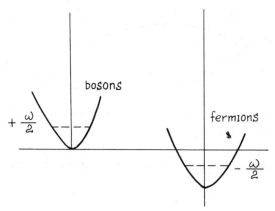

Figure 5.5: Vacuum energy of bosons and leptons. The ground state energies differ by sign.

exists a deep symmetry between fermions and bosons called supersymmetry. If this is true then there exists a one-to-one correspondence between every bosonic and fermionic state. Experiment shows, however, that this symmetry is not exact since masses of bosons and fermions are different. Therefore vacuum energy is not exactly cancelled out even if there are an equal number of fermionic and bosonic states. The contributions to vacuum energy which diverge as the fourth and even as the second power of the integration upper bound do cancel out and the nonvanishing remnant is proportional to the fourth power of mass difference i.e. to m_{SUSY}^4, where m_{SUSY} is a mass parameter describing the scale of supersymmetry breaking. According to existing data $m_{SUSY} > 10^3$ GeV which gives $\rho_{vac} > 10^{12} \text{GeV}^4$ in some (!) contradiction with the bound (5.18).

Apart from that there are plenty of phase transitions in the primordial plasma in the course of the expansion and cooling down of the Universe. There is, for example, the transition from quark-gluon plasma to hadron matter, or from symmetric to nonsymmetric electroweak phase. Each of these phase transitions is accompanied by a change in the energy momentum tensor which is equal to $\delta\rho g_{\mu\nu}$. Evidently this is the change in vacuum energy.

Let us explain this in more detail. Let a system be described by the potential presented in figure 5.6, and let the wave function

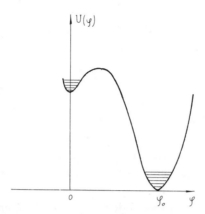

Figure 5.6: The effective potential of a system with false vacuum at $\phi = 0$ and true vacuum at $\phi = \phi_0$.

be initially localized around $\phi = 0$. If the tunnelling probability is small enough then the point $\phi = 0$ to a good approximation can be considered as the vacuum state of the model. It is called the false vacuum. One can discuss excitations over this vacuum and formulate the usual field theoretic model neglecting the existence of the real vacuum at $\phi = \phi_0$. If the transition probability is small one might live over the false vacuum for a long time not suspecting of being on a "powder-magazine". The tunnelling probability being non zero, sooner or later the system must go to $\phi = \phi_0$. This transition is accompanied by the energy release $\Delta\rho = U(0) - U(\phi_0)$. $\Delta\rho$ is equal to $10^{-4}\mathrm{GeV}^4$ for the quark-hadron phase transition, to $10^8\mathrm{GeV}^4$ for the electroweak one and to $10^{60}\mathrm{GeV}^4$ for analogous grand unification phase transition (these transitions will be discussed in chapter 6). There may still remain some doubt regarding the earlier phase transitions but the quark-hadron phase transition is absolutely sure. In particular, the existence of gluon condensate with energy momentum $T_{\mu\nu} = g_{\mu\nu}(0.1\mathrm{GeV})^4$ is practically established by experiment because the observed hadron properties well coincide with those which were found theoretically under the assumption of the gluon condensate.

The value of vacuum energy density after all the phase transitions in the Universe is amazingly close to zero. The idea that the initial value of the Λ-term was fine tuned with the accuracy of about one hundred decimal places to ensure cancellation, is clearly crazy. There must exist a dynamical model automatically providing the compensation of the Λ-term. This compensation does not have to be instant and its natural time scale might be that of expansion of the Universe. Therefore some deviations from the Friedman model could exist at any stage and even now. Still, no concrete predictions can be made because, as yet, there exists no absolutely satisfying model of Λ-term compensation. This problem is a central one in both cosmology and particle physics and provides a serious challenge to theoreticians.

We would like to quote a joke by I.Ya.Pomerantschuk to end this chapter :All physics is the physics of vacuum - the first part is "Pumps and manometers" and the second one is "Quantum theory of vacuum".

Chapter 6

Scalar fields in cosmology

The scalar field was introduced to physics by Yukawa in 1935. Very soon quanta of this field (π mesons) were found. In the sixties it appeared that they are not fundamental but consist of quarks. The scalar field was banished from the list of fundamental fields but not for a long time. Gauge theories with spontaneously broken symmetries demanded the existence of scalar fields in order to ensure renormalizability.

Supersymmetry adds another argument in favour of the existence of a scalar field. The great expectations to find a unified field theory are based on supersymmetry. Supersymmetry binds fields with different spins and therefore makes another step on the path to unification of fields and interactions. Particles with different spins are believed to be different states of a fundamental field. The existence of fields with spins 1, 1/2 and 2 implies the existence of fields with spin 0 and 3/2.

A scalar field can naturally lead to the equation of state $p = -\rho$ which is very interesting from the cosmological point of view.

6.1   Scalar field in flat spacetime

The Lagrangian of a complex scalar field has the following form:

$$L = |\partial_\mu \phi|^2 - U(\phi) \qquad (6.1)$$

119

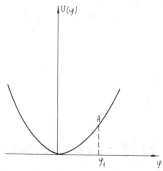

Figure 6.1: Potential of free scalar field.

where the potential $U(\phi)$, in the case of noninteracting particles, is:

$$U_0(\phi) = m^2|\phi|^2 \tag{6.2}$$

This expression differs by a factor $1/2$ from equation (3.6) which was written for the real field. Rewriting the Lagrangian (6.1) in terms of two real fields ϕ_1 and ϕ_2 defined by the expression $\phi = (\phi_1 + i\phi_2)/\sqrt{2}$ one gets the sum of the two Lagrangians (3.6). The potential $U_0(\phi)$ is shown in figure 6.1 (In fact it is only a longitudinal section of the potential. The full potential can be obtained by a rotation around U axis). Let us note that the Lagrangian (6.1) with potential (6.2) is invariant with respect to the phase transformations:

$$\phi \rightarrow \phi \exp(i\alpha) \tag{6.3}$$

The potential of the Higgs field presented in Fig. 4.12 can be written as:

$$U_H(\phi) = \frac{1}{4}\lambda\left(\phi_0^2 - |\phi|^2\right)^2 \tag{6.4}$$

It is also invariant with respect to transformations (6.3) like every potential $U(|\phi|)$.

The point $\phi = 0$ is not a point of stable equilibrium of the potential (6.4). The stable equilibrium points form a line in the complex ϕ plane:

$$\phi = \phi_0 \exp(i\sigma_0) \tag{6.5}$$

with arbitrary phase σ_0. Expansion of the Lagrangian around one of the stable points can be done by making the substitution $\phi = \phi_0 + \chi \exp(i\sigma)$. Thus:

$$L = (\partial_\mu \chi)^2 + (\phi_0 + \chi)^2 (\partial_\mu \sigma)^2 - \lambda(\phi_0^2 \chi^2 + \phi_0 \chi^3 + \chi^4/4) \tag{6.6}$$

This Lagrangian describes two fields: the first one, χ, is massive with mass $m^2 = \lambda \phi_0^2$ and self interaction $1/4\lambda(\chi^4 + 4\phi_0\chi^3)$, and the second one σ_0 is massless. The invariance with respect to transformations (6.3) is now absent. The symmetry has been spontaneously broken. The appearance of a massless field σ is a consequence of the general theorem by Goldstone. If, however, a vector gauge field interacting with ϕ is added to the Lagrangian then no massless field appears, and the corresponding degree of freedom becomes the longitudinal component of the vector field. Exactly the same happens in the case of the electroweak interaction.

We are particularly interested in the energy-momentum tensor of the field ϕ as it is a source of the gravitational field. Let us write it down for the simple case of a real scalar field with Lagrangian (3.6). The energy-momentum tensor is defined as a functional derivative of the action with respect to the metric:

$$T_{\mu\nu} = \frac{\delta}{\delta g^{\mu\nu}} \int d^4x \sqrt{g}[g^{\alpha\beta}\partial_\alpha\phi\partial_\beta\phi - U(\phi)] =$$

$$= \partial_\mu\phi\partial_\nu\phi - \frac{1}{2}g_{\mu\nu}[g^{\alpha\beta}\partial_\alpha\phi\partial_\beta\phi - U(\phi)] \tag{6.7}$$

In the case of a homogeneous field depending on time only, $T_{\mu\nu}$ has the form:

$$T_{00} \equiv \rho = (\dot{\phi}^2 + m^2\phi^2)/2 \tag{6.8}$$

$$T_{ij} = p\delta_{ij} = \delta_{ij}(\dot{\phi}^2 - m^2\phi^2)/2 \tag{6.9}$$

It follows from these expressions that if at some moment t a homogeneous field $\phi(t)$ satisfies two conditions: $\dot{\phi} = 0$ and $\phi \neq 0$ the relation $p = -\rho$ is fulfilled at that moment.

For potential (6.2) these conditions may be satisfied only in discrete moments of time since the equations of motion imply $\ddot{\phi} \neq 0$ if $\dot{\phi} = 0$ and $\phi \neq 0$. If, however, $\dot{\phi}$ changes slowly (i.e. $\ddot{\phi}$ is small enough) then the relation $p = -\rho$ holds for a long time. In the case of Higgs field the conditions $\dot{\phi} = 0$ and $U(\phi) \neq 0$ may be valid independently of time.

We would like to demonstrate once more Lorentz invariance of the relation $p = -\rho$ (see sec. 1 chapter 5). Note, that the vanishing of all derivatives of ϕ is invariant with respect to change of coordinate system and, in particular, to Lorentz transformations. Since ϕ is a scalar field it is not affected by Lorentz transformations. The gradient of a scalar field is a four dimensional vector. The values of vector components are changed under these transformations except for a special case when all the components are zero. The zero vector is the only Lorentz invariant vector. The invariance of field ϕ and its vanishing derivatives implies the invariance of the relation $p = -\rho$.

Now we shall discuss the physical meaning of the initial conditions $\dot{\phi} = 0$ and $\phi \neq 0$. The equation of motion of a homogeneous field ϕ is:

$$\frac{d^2\phi}{dt^2} = -\frac{\partial U(\phi)}{\partial \phi} \tag{6.10}$$

This is the second order differential equation which is solved with two arbitrary initial conditions. Our choice is $\dot{\phi} = 0$ and $\phi = \phi_1 \neq 0$

Equation (6.10) is identical to equation of motion of a massive point particle in classical mechanics. The field ϕ plays the role of the coordinate of the particle and $U(\phi)$ is the potential, in which the particle moves. The potential $U = m^2\phi^2/2$ is the potential of a harmonic oscillator with frequency m. The initial conditions $\dot{\phi} = 0$ an $\phi = \phi_1$ mean that the particle was initially shifted to the distance ϕ_1 from the equilibrium point and did not move. This is illustrated by figure 6.1. It can be easily seen that the initial state is not stable. As soon as the external forces cease to hold the particle

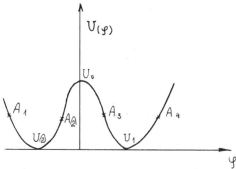

Figure 6.2: Potential in the Higgs model.

at point A, it starts to roll down. The motion of the particle or in other words the time dependence of field ϕ is described by the simple oscillator formula:

$$\phi(t) = \phi_0 \cos(mt + \psi)$$

The initial conditions determine the phase of the oscillations ψ. The state $\dot{\phi} = 0$ and $\phi = \phi_1$ is repeated after every period and $\dot{\phi}$ vanishes after every half period, and so the relation $p = -\rho$ is valid for a moment after every half period.

The Higgs potential, shown in figure 6.2, is more complicated. It has three equilibrium points. One of them (U_0) is unstable and the other two are stable $(U_1$ and $U_2)$. If the system is initially at one of the points A1, A2, A3, A4 its evolution will be the same as described above. Now let us chose $\phi = 0$ (the unstable equilibrium) as the initial state. The field cannot remain in this state for an infinitely long time, since sooner or later quantum fluctuations pull the field out of the equilibrium. It is noteworthy that the state $\phi = 0$ corresponds, according to equation (6.4), to a non-vanishing value of the potential:

$$U(0) = \lambda\phi^4/4 \qquad (6.11)$$

It means that the relation $p = -\rho$ holds if ϕ is at the top of the potential barrier. There is one more interesting property of a scalar field. At the moment when $\phi = 0$ and $\dot{\phi} \neq 0$, which corresponds to a passage through the equilibrium state in figure 6.1, the maximally stiff equation of state:

$$p = +\epsilon \tag{6.12}$$

is realized. In a medium with such an equation of state the speed of sound is maximal i.e. it would reach the speed of light. This equation was proposed 20 years ago in vector field theory by Ya.B.Zeldovich but as we have just seen it can be realized for a scalar field.

The notion of an equation of state is rather ambiguous in this case since the system is not relaxed and p is not determined by ρ. The relation between p and ρ changes with time and there is no dynamical equation which would describe the evolution of p and ρ, once their initial values are given. For that the equation of motion of the field ϕ is now necessary. For the free field case this is, as we already know, the Klein-Gordon equation:

$$(\partial^2 + m^2)\phi \equiv (\partial_t^2 - \nabla^2 + m^2)\phi = 0 \tag{6.13}$$

This equation has the flat waves solutions:

$$\phi = \phi_0 \exp(-i\omega t + ikr) \tag{6.14}$$

where:

$$\omega^2 = k^2 + m^2 \tag{6.15}$$

The homogeneous solution corresponding to vanishing momentum is:

$$\phi(t) = \phi_0 \exp\left(\pm i(mt + \psi)\right)$$

or

$$\phi(t) = \phi_0 \cos(mt + \psi) \tag{6.16}$$

The wave function oscillation frequency is not zero even if the particle is at rest, since the lowest energy of a particle is its mass (or mc^2 if $c \neq 1$). Let us calculate the pressure and energy density corresponding to the solution (6.16):

$$\rho = \frac{\dot{\phi}^2}{2} + \frac{m^2\phi^2}{2} = \frac{m^2\phi_0^2}{2}$$

$$(6.17)$$

$$p = \frac{\dot{\phi}^2}{2} - \frac{m^2\phi^2}{2} = \frac{m^2\phi_0^2 \cos(2mt + 2\psi)}{2}$$

Energy density as one expects does not depend on time. What is rather surprising is that the pressure is not zero though the particles are at rest but time averaged pressure does vanish. If the field ϕ is a superposition of states (6.16) with random phases i.e. it represents a set of non-correlated particles then the pressure vanishes after phase averaging. Thus we have obtained a natural result that noncoherent superposition of particles at rest has zero pressure. Therefore, it satisfies the dust equation of state $p = 0$.

The field ϕ can be formed by a coherent superposition of quanta. This is a classical field and the pressure is a non-zero oscillating function of time. The mean pressure, however, is zero. This is true if the potential is $U(\phi) = m^2\phi^2/2$ but in the case of other potentials it does not have to be zero.

If the characteristic time scale is small in comparison with m^{-1} the coherent state of ϕ can be approximately described by the equation of state $p = p(\rho)$ which may change from the maximally stiff $p = \rho$ to the vacuum one $p = -\rho$. Sometimes it is a convenient idealization of a more complicated real state of the system under consideration. The point is that strictly speaking there is no equation of state which would allow one to express $p(\phi, \dot{\phi})$ through $\rho(\phi, \dot{\phi})$. But if the field variation is slow enough the notion of equation of state makes sense. The fact that the effective equation of state changes from $p = \rho$ to $p = -\rho$ leads to the idea that the Universe may expand exponentially during the period, when the relation $p = -\rho$ holds. To make a more rigorous statement we have to consider the equation of motion of ϕ in the expanding Universe.

The important question, from the point of view of cosmological applications, is how long could the relation $p = -\rho$ be valid. It will later be shown that a friction force appears in field equations (6.10) due to expansion of the Universe. This force slows down the movement of field ϕ to the equilibrium. Therefore the inflation could last long enough.

Let us note in advance that the expansion of the Universe by itself does not alter the relation $p = -\rho$ but if ϕ is not constant the inner forces change p and ρ and also the relation between them.

6.2   Scalar field in the expanding universe

We have discussed above matter fields in flat spacetime when the gravitational field is negligible. Some qualitative properties of the fields have been established. Now we shall take gravity into account. This means that the fields must be considered in the curved spacetime. In a cosmological situation the metric of such a space is nonstationary. On the other hand the metric is determined by the energy momentum tensor of matter fields. Hence the self-consistent equations of motion of matter and gravitational fields should be solved. As a preliminary example we consider scalar field equations in a curved spacetime with fixed, homogeneous metric. This approximation is meaningful when the gravitational field is determined by other matter fields and the contribution of ϕ to the total energy momentum tensor is negligible.

Equations of motion in a curved spacetime are obtained through the standard procedure of changing ordinary derivatives ∂_μ to covariant ones $D_\mu = \partial_\mu + \Gamma_\mu$, where Γ_μ depends on the object to which D_μ is applied. The meaning of this change is the following. Differentiating in flat spacetime in cartesian coordinates is a tensor operation. If ϕ is a scalar then $\partial_\mu\phi$ is a vector , if A_μ is a vector then $\partial_\mu A_\nu$ is a tensor, etc. This is not true in curved spacetime nor even in a flat spacetime with curved, for example spherical, coordinates. The derivative operator leads to an object which no longer is a tensor. The point is that differentiation includes finding a difference of functions in two close but different points. Tensors in different points are transformed differently in curved spacetime and therefore their difference is not a tensor. The above presented modification of derivative is introduced in order to compensate this effect. It follows, that a covariant derivative of a scalar quantity is identical to the ordinary one. In case of a vector the covariant derivative is:

$$D_\mu V_\nu = \partial_\mu V_\nu - \Gamma^\lambda_{\mu\nu} V_\lambda$$

where $\Gamma^\lambda_{\mu\nu} = 1/2 g^{\lambda\sigma}(\partial_\nu g_{\mu\sigma} + \partial_\mu g_{\nu\sigma} - \partial_\sigma g_{\mu\nu})$ is the Christoffel's symbol. When applied to a tensor of an arbitrary rank quantities $\Gamma^\lambda_{\mu\nu}$ act successively on each index as follows:

$$D_\mu T_{\nu\gamma...} = \partial_\mu T_{\nu\gamma...} - \Gamma^\lambda_{\mu\nu} T_{\lambda\gamma...} - \Gamma^\lambda_{\mu\gamma} T_{\nu\lambda...} - ...$$

It can be easily checked that $D_\mu g_{\alpha\beta} = 0$.

The scalar field equation of motion in the background of a homogeneous nonstationary metric is:

$$\ddot{\phi} - \frac{1}{a^2}\Delta\phi + m^2\phi + 3H\dot{\phi} = 0 \qquad (6.18)$$

where $H = \dot{a}/a$ is the Hubble parameter. We are interested here in homogeneous (i.e. not depending on spatial coordinates) solutions only. This leads to the oscillator equation with variable friction:

$$\ddot{\phi} + 3H\dot{\phi} + m^2\phi = 0 \qquad (6.19)$$

The friction is generally not constant since the Hubble parameter may depend on time.

We should, however,solve a self consistent problem when the evolution of $H(t)$ is determined by scalar field ϕ. Let us first discuss a simpler case when $H = $ const which corresponds to the de Sitter spacetime.Physically this means that the energy-momentum tensor is determined by other fields (including maybe other scalars) which ensure the equation of state is $p = -\rho$ and correspondingly the de Sitter regime of expansion, $H = $ const.

We shall look for solutions of (6.19) in the form:

$$\phi(t) = \phi_0 \exp(\lambda t)$$

where λ is a complex number satisfying the equation:

$$\lambda^2 + 3H\lambda + m^2 = 0 \tag{6.20}$$

It has the following roots:

$$\lambda = -\frac{3H}{2} \pm \sqrt{\frac{9}{4}H^2 - m^2} \tag{6.21}$$

When $m > 3H/2$ the solution is close to that in a stationary homogeneous background i.e. to $\exp(imt)$. The only difference is that the amplitude decreases adiabatically with the decrement $3H/2$.

A more interesting case is when $m \ll H$. This means that the background changes much faster than the field ϕ. Roots (6.21) can be approximately written as:

$$\lambda_1 = -3H; \quad \lambda_2 = -m^2/3H \tag{6.22}$$

And so there are two solutions of equation (6.19):

$$\phi_1(t) = \phi_1 \exp(-3Ht)$$

and

$$\phi_2(t) = \phi_2 \exp(-m^2 t/3H)$$

The first solution decreases much faster than the second one since $3H \gg m^2/3H$. Hence, after some time from the initial superposition:

$$\phi(t) = \phi_1 \exp(-3Ht) + \phi_2 \exp(-m^2 t/3H) \tag{6.23}$$

only the second term survives. The only exception is the case when $\phi_2 = 0$.

The energy density and pressure corresponding to solution (6.23) are:

$$\rho = \frac{1}{2}\left(3H\phi_1 e^{-3Ht} + \frac{m^2}{3H}\phi_2 e^{\frac{-m^2 t}{3H}}\right)^2 + \frac{m^2}{2}\left(\phi_1 e^{-3Ht} + \phi_2 e^{\frac{-m^2 t}{3H}}\right)^2 \rightarrow$$

$$\rightarrow \frac{m^2}{2}\phi_2^2 e^{\frac{-2m^2 t}{3H}}\left(1 + \frac{m^2}{9H^2}\right)$$

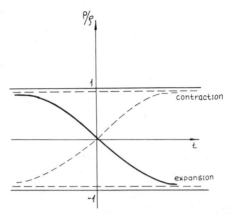

Figure 6.3: Evolution of the p/ρ-ratio in the expanding (solid line) and contracting (dotted line) Universe. The asymptotic values differ from unity by $2m^2/9H^2$ when $m \ll H$.

$$p = \frac{1}{2}\left(3H\phi_1 e^{-3Ht} + \frac{m^2}{3H}\phi_2 e^{\frac{-m^2 t}{3H}}\right) - \frac{m^2}{2}\left(\phi_1 e^{-3Ht} + \phi_2 e^{\frac{-m^2 t}{3H}}\right)^2 \rightarrow$$

$$\rightarrow -\frac{m^2}{2}\phi_2^2 e^{\frac{-2m^2 t}{3H}}\left(1 - \frac{m^2}{9H^2}\right)$$

One can see that the relation between p and ρ exponentially tends to $p = -\rho$ with the accuracy of the order of $(m/3H)^2$. Vice versa, when the Universe shrinks (which corresponds to the change of sign of H) only the first term in (6.23) remains, and the equation of state approaches the maximally rigid one $p = \rho$.

In the degenerate cases when $\phi_1 = 0$ or $\phi_2 = 0$ the relations $p = -\rho$ or $p = \rho$ respectively are valid independently of time. Generally, the relation between p and ρ changes with time and approaches $p = -\rho$ when the Universe expands and $p = +\rho$ when it contracts. The ratio p/ρ versus time in both cases is depicted in figure 6.3. The contraction is obtained from the expansion by the transformation $t \rightarrow -t$.

Horizontal lines correspond to the degenerate cases $\phi_1 = 0$ and $\phi_2 = 0$. The solid line corresponds to the expansion and the dashed one corresponds to the contraction. The position and the shape of these curves depend on the initial conditions but the asymptotic limit is always the same and is determined by the evolution of the Universe.

In other words the relation $p = +\rho$ is unstable when the Universe expands while $p = -\rho$ is unstable when it contracts.

It resembles the Le Chatelet-Braun principle in thermodynamics. In equilibrium pressure is a function of density $p = p(\rho)$ and entropy is constant. Hence for any closed path in the phase plane p remains the same. In non-equilibrium processes the pressure becomes larger than the equilibrium pressure when the gas is compressed. An extra pressure is developed in order to resist the compression. On the other hand during the expansion the gas resists it and the pressure is lower than the equilibrium one. Hence in real gases entropy increases after going round a closed circle, $\Delta S > 0$. One should realize that the equation of state must contain corrections of the form:

$$p = p(\rho, S) + a\dot{\rho} = p(\rho, S) - b\dot{V}, \quad \text{where} \quad a > 0 \quad \text{and} \quad b > 0.$$

These corrections are in accordance with the Le Chatelet-Braun principle. The evolution of the field ϕ is analogous to the behaviour of pressure in this thermodynamical example.

Let us now analyze the self consistent problem. First, a simplified version will be discussed. It will be assumed that the second derivative of ϕ with respect to time, $\ddot{\phi}$, is much smaller than $H\dot{\phi}$ and $m^2\phi$. It will also be assumed that the field ϕ and the background metric change slowly i.e. $\dot{\phi} < m\phi$ and $\dot{H} \ll H^2$. The self consistent equations for the field ϕ and the scale factor are:

$$\frac{H^2}{2} = \frac{4\pi G\rho}{3} \quad \text{or} \quad H = \frac{m\phi}{m_{Pl}}\sqrt{\frac{4\pi}{3}} \tag{6.24}$$

In this simplified case the solution can be easily written down:

$$\phi = \phi_0 - m\, m_{Pl}\, t/\sqrt{12\pi} \tag{6.25}$$

This formula holds for $0 < t < \sqrt{12\pi}\phi_0/mm_{Pl}$. In this time interval the Hubble parameter is:

$$H = \frac{\sqrt{4\pi/3}\, m}{m_{Pl}} \left(\phi_0 - \frac{m\, m_{Pl}\, t}{\sqrt{12\pi}} \right) \qquad (6.26)$$

Once the expression for $H(t)$ is known one can calculate the most important inflationary parameter, the degree of expansion, as a function of field ϕ. The degree of expansion for the ideal de Sitter solution is equal to $H\Delta t$ where Δt is the duration of de Sitter stage. In the case of approximate de Sitter expansion when H is a function of time it is given by $\int H dt$. In the model presently considered:

$$\int H dt = 2\pi(\phi_0/m_{Pl})^2$$

If e.g. $\phi_0 = 4m_{Pl}$ then the inflation will last long enough i.e. the degree of expansion will be about 100 and $a_{final} = a_0 \exp(100) = 10^{43} a_0$.

Let us now consider the self consistent problem more rigorously. It will be assumed that the expansion rate is determined by the scalar field and that the metric as well as the field are homogeneous. We shall discuss the problem of how probable is inflation or in other words whether the solutions resulting in inflation constitute a small part of all the possible solutions. It will be proved that this is not the case. Inflation is realized for a very wide class of initial conditions. Therefore we believe that exponential expansion arises naturally regardless of the initial state if there exists a scalar field in the early Universe. To prove this we shall use eq. (6.18) as above but no assumptions will be made about the rate of variation of ϕ and $a(t)$. Our consideration is based on the paper by Belinsky, Grishchuk, Khalatnikov and Zeldovich (1985).

The equations of motion of interacting scalar and gravitational fields are:

$$\ddot{\phi} + 3H\dot{\phi} + U'(\phi) = 0 \qquad (6.27)$$

$$H^2 + \frac{k}{a^2} = \frac{1}{6}G(m^2\phi^2 + \dot{\phi}^2) \qquad (6.28)$$

This system of equations does not possess the exact De Sitter solution. The potential vanishes in the only stable point $\phi = 0$ and

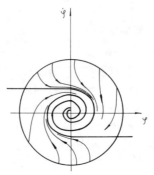

Figure 6.4: Phase trajectories of field ϕ in the expanding Universe.

so the relation $p = -\rho$ does not hold. When, however, $\phi \neq 0$ then the movement to the stable point can be slow enough so that an approximate de Sitter solution is valid. From the point of view of applications an approximate solution is as good as the exact.

The properties of equations (6.27) and (6.28) are best illustrated in the phase space i.e. in the space of ϕ, $\dot{\phi}$ and H. Trajectories in this space describe possible evolution of $\phi(t)$ and a(t). In the case of flat spacetime, i.e. for $k = 0$, the phase space is simplified, it becomes two-dimensional (figure 6.4). Each curve on the diagram describes a possible solution.

There are two separatrixes dividing two families of the solutions. They begin on the circle with radius unity and go parallel to axis OX and then spiral around the centre. Different points on this circle correspond to different initial values of ϕ and $\dot{\phi}$. The scale was chosen so that the radius would be unity. The infinite time corresponds to asymptotic movement of ϕ and $\dot{\phi}$ to zero. The initial conditions should be set on this circle, which corresponds to $t = t_{Pl}$. But at that time quantum gravitational effects are essential and the system of equations (6.27) and (6.28) is not valid. This difficulty can be avoided if one fixes the initial conditions inside the unit circle i.e. for the $t > t_{Pl}$. If the initial values of ϕ and $\dot{\phi}$ lie on a separatrix then the evolution is closer to the de Sitter one than any other. In other words, if at $t \approx t_{Pl}$ the initial conditions were $\dot{\phi} = 0$ and $\phi = 1$ then the Universe underwent a long inflationary stage. But even if the initial point was outside the separatrix a considerable inflation mostly also took place.

Starting from almost any point on the circle the trajectory quickly approaches one of the separatrixes. Until that moment the Universe evolves in accordance with the equation $p = +\rho$. Later the trajectory moves along one of the separatrixes and the effective equation of state becomes $p = -\rho$, as required for inflation. During that time the scale factor increases more than 10^{30} times if $H\tau > 70$.

After the inflation is over the trajectory starts to spiral around the centre just as the separatrixes do. The trajectories corresponding to the Universe without or with very short ($H\tau < 70$) inflation begin in the small region $\phi \approx 0$ around OY-axis. The size of this region, compared to the size of the circle is about 10^{-6}. Thus, the scalar field leads to inflation with the probability equal to $1 - 10^{-6}$.

In this model inflation starts just after the Universe is born. It can be said that the Universe is cool initially and heats after inflation. There are also models in which initially hot Universe cools down during the inflation and is later reheated. Such a model will be discussed in the next section.

6.3 Inflation and the Higgs field

Historically, the first inflationary model based on scalar field dynamics used a different mechanism than that described in section 2. It utilized special properties of the Higgs potential (6.4) and in particular its temperature dependence which was not accounted for in expression (6.4). Qualitatively this dependence can be found in the following way.

Let ϕ be a massive scalar field with self interaction $\lambda\phi^4$ i.e its potential is:

$$U = m_0^2\phi^2 + \lambda\phi^4/4 \tag{6.29}$$

Since the Lagrangian of the theory is symmetric with respect to the transformation $\phi \to -\phi$ the expectation value of the field in the ground state vanishes, $\langle\phi\rangle = 0$, exactly like the expectation value of coordinate in a quantum mechanical oscillator, $\langle x \rangle = 0$. It is, however, known that $\langle x^2 \rangle$ is not zero. The expectation value of the field squared is also not zero because of quantum effects:

$$\langle \phi^2 \rangle \neq 0$$

In the field theory this value is infinite because the system has infinitely many degrees of freedom. The procedure of renormalization mentioned in sec.7 of chapter 4 takes care of this infinity, which is absorbed by vacuum energy and the particle mass. The contribution to the vacuum energy is:

$$\rho_{vac} = m^2 \langle \phi^2 \rangle_0 + \lambda \langle \phi^4 \rangle_0$$

Quantum corrections to the mass arise in the following way:

$$\lambda \phi^4 \rightarrow \lambda \langle \phi^2 \rangle_0 \phi^2$$

i.e. $\delta m^2 = \lambda \langle \phi^2 \rangle_0$ and the physical mass is $m_0^2 + \lambda \langle \phi^2 \rangle_0$, the index 0 meaning zero temperature.

What happens when the system is in contact with a heat bath with temperature $T \neq 0$? Having in mind the quantum oscillator we can conclude that $\delta m^2(T) = \lambda(\langle \phi^2 \rangle_0 + bT^2)$, where b is a constant. Thus the effective potential [1] of the field at nonzero temperature is:

$$U(\phi, T) = m^2 \phi^2 + bT^2 \phi^2 + \lambda \phi^4 / 4 \qquad (6.30)$$

Analogously, the Higgs potential at nonzero temperature is modified in the following way:

$$U(\phi, T) = (bT^2 - \frac{\lambda}{2}\phi_0^2)\phi^2 + \frac{\lambda}{4}\phi^4 + \frac{\lambda}{4}\phi_0^2 \qquad (6.31)$$

$U(\phi, T)$ for different values of T is presented in figure 6.5. At high temperatures the potential has one minimum at $\phi = 0$. There is no field condensate i.e. $\langle \phi \rangle = 0$ and the system is in a symmetric state. If there are gauge fields interacting with ϕ their mass without temperature corrections is zero.

---

[1]Strictly speaking the free energy F should be considered instead of the effective potential since equilibrium is determined by the minimum of F.

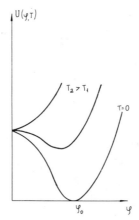

Figure 6.5: Effective potential of scalar field at different temperatures.

At smaller temperatures an additional minimum in the potential $U(\phi, t)$ appears. While the temperature goes down this minimum becomes deeper than that at $\phi = 0$ so the point $\phi = 0$ becomes unstable and the system evolves to the state with nonvanishing average value of the field i.e. $\langle \phi \rangle \neq 0$. The symmetry is spontaneously broken and gauge fields acquire mass according to the mechanism discussed in section 8 in chapter 4. This phenomenon was discovered by D.A.Kirzhnits (1972) and D.A.Kirznits, A.D.Linde (1972).

The time, the system remains in the metastable state $\phi = 0$, depends on λ and ϕ_0. If :

$$-m_1^2 = -\left.\frac{\partial^2 U}{\partial \phi^2}\right|_{\phi=0, T=0} > H^2 \frac{8\pi}{3} \frac{\lambda \phi_0^4}{12 m_{Pl}^2}$$

then the system goes from $\langle \phi \rangle = 0$ to $\langle \phi \rangle = \phi_0$ without any delay. This is a phase transition of the second order and it is of no particular interest to us. Inflation is realized if $m_1^2 < 0$ and $|m_1^2| < H^2$ or $m_1^2 > 0$ i.e. there is a local minimum at the point $\phi = 0$. The system remains for a long time in the metastable phase (exponentially long in the case $m_1^2 > 0$). The gradient terms $\partial_\mu \phi$ or in other words, the surface energy of the bubbles of new phase, prevent the transition to the phase with $\langle \phi \rangle \neq 0$. If $m_1^2 > 0$ the phase transition resembles formation of vapour bubbles in a heated fluid. The critical size of bubbles of new phase should be big enough so that

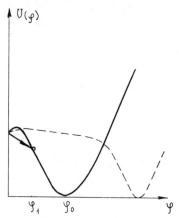

Figure 6.6: Scalar field tunnelling from $\phi = 0$ to $\phi = \phi_0$ leading to phase transition of the first order.

the volume energy be larger than the surface energy. This leads to a small probability of phase transition and the field sticks near $\phi = 0$ for a long time (figure 6.6). At this period the ordinary matter density decreases as T^4 while the energy of the field ϕ remains constant:

$$\rho_0 \equiv \rho(\phi = 0) = \lambda \phi_0^4/4 \qquad (6.32)$$

Using eqs.(6.7) and (6.11) one can prove that $T_{\mu\nu} = (\lambda\phi_0^4/4)g_{\mu\nu}$. This energy momentum tensor acts effectively like a cosmological constant. If the system remains at the point $\phi = 0$ long enough then ρ_0 becomes greater than the energy density of all other forms of matter and the Universe starts to expand exponentially:

$$a(t) \approx \exp(Ht), \quad H = \frac{\sqrt{\lambda}\,\phi_0^2}{2\,m_{Pl}} \qquad (6.33)$$

Soon the energy density of other forms of matter becomes negligible (energy density of relativistic matter scales as $\exp(-4Ht)$ and that of nonrelativistic matter scales as $\exp(-3Ht)$). The Universe becomes as isotropic and homogeneous as vacuum can be and its expansion is determined by the vacuum-like energy (6.32). Note that we have implicitly assumed that there is a nonvanishing cosmological constant $\Lambda = \lambda\phi_0^4/32\pi m_{Pl}^2$ in the symmetric phase and that it is strictly cancelled out by the energy of the condensate when $\langle \phi \rangle = \phi_0$. This is an example of fine tuning of parameters which has been considered in section 4 of chapter 5.

As it has already been noted this phase transition resembles boiling of water and formation of bubbles in the overheated fluid. A real fluid cannot remain long in the overheated state since there are a lot of impurities in it which serve as seeds of the new phase. If, however, no such seeds are present the system may remain in the metastable phase for a long time. If this is true for the field ϕ the delay of the phase transition can be large enough so that the condition $H\tau > 70$ is fulfilled.

Formation of the bubbles of new phase corresponds to a tunnelling transition from point ϕ_0 to point ϕ_1 (see figure 6.6). The estimates of transition probability prove that the inflation can be sufficiently long. After tunnelling the field approaches a stable equilibrium point according to the equation:

$$\ddot{\phi} + 3H\dot{\phi} + U'(\phi) = 0 \qquad (6.34)$$

Solutions of this equation are very similar to those of equation (6.19) where $U = m^2\phi^2/2$ since the Higgs potential is close to quadratic around ϕ_0.

This scenario was proposed by A.Guth in 1981. This model, however, suffers the following shortcoming. The field ϕ quickly approaches its equilibrium. Thus, the bubble size was very small and the visible part of Universe should contain many bubbles. This leads to large inhomogeneities because the energy density contrast inside a bubble and on its border is huge. Linde (1982) and Albrecht and Steinhardt (1982) modified this scenario getting rid of the trouble. The potential they proposed is very flat around ϕ_1 so the rate of variation of the field ϕ is small in comparison with the rate of expansion of the Universe (see figure 6.6 dashed line):

$$\dot{\phi}/\phi \ll H_0 \qquad (6.35)$$

Thus we come to the case considered in section 2. The expansion of the Universe for $\phi = 0$ and $\phi = \phi_1$ differ only slightly. This means that the bubble of new phase with $\langle \phi \rangle = \phi_1$ expands exponentially

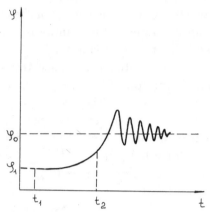

Figure 6.7: Time variation of inflaton field. The point t_1 corresponds to the moment of tunnelling. The field changes slowly and the Universe keeps on expanding exponentially between t_1 and t_2. The field starts to oscillate after t_2. This oscillations lead to particle creation and reheating.

and its border goes far outside the present day horizon. Our Universe, in this model, is a small inner part of a gigantic bubble, that is why it looks isotropic and homogeneous.

6.4   End of inflation and reheating of the universe

At this moment the reader might ask the question: what does one get with this inflationary model? The problems of flatness, horizon and homogeneity are of course solved but at what price? There is nothing left in the Universe except for the condensate of scalar field ϕ. This dull picture does not resemble the Universe we live in so either the model is wrong or there must be a way to describe the present Universe in this approach. Fortunately the second possibility can be realized.

The time dependence of the field ϕ is governed by equation (6.34). Figure 6.7 illustrates it qualitatively.

The bubbles of new phase are formed predominantly in the state where the potential of the field ϕ is flat (see figure 6.6) and ϕ

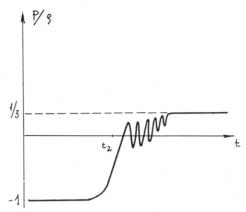

Figure 6.8: Evolution of p/ρ ratio in the inflationary model. The inflation ends at t_2.

grows very slowly. As the field approaches ϕ_0 the potential becomes steeper and the field starts to move faster. Having reached the equilibrium point, the field begins to oscillate around it. The oscillations are damped by friction caused by expansion of the Universe ($3H\dot{\phi}$ term) and by particle production which is not taken into account in equation (6.34).

It is known that an oscillating electric field can produce electron-positron pairs and that the production rate is not suppressed if the frequency is high $\omega > m_e$. If the field ϕ interacts with any elementary particles it can, like an electric field, create these particles. The field oscillation frequency in our model is equal to its mass, m_ϕ. In grand unified theories $m_\phi > 10^{14} - 10^{15}$GeV. The oscillating field creates particles with mass $m < m_\phi$. Energy of the field is transformed into energy of these particles. If the expansion of the Universe is slower than the rate of particle production and the rate of reactions between them then these particles are thermalized and we come to the standard cosmological model. Some deviations from thermodynamical equilibrium are expected at the initial stage and this helps to solve the problem of baryon asymmetry of the Universe (Dolgov, Linde 1982).

Therefore the scalar field not only causes inflation but produces all the matter in the Universe.

In this scenario, there was a hot stage before inflation. Then the Universe exponentially quickly cooled down and after that one more heating up occurred. The thermal history of the Universe in terms of the relation between p and ρ is presented in figure 6.8.

Chapter 7

Inflation and vacuum polarization

Quantum field theory has drastically changed our views, not only on particles interactions, but also on properties of empty space, vacuum. It might seem, that nothing can be simpler than vacuum. We know now that it is a very complicated object. Vacuum properties cause quark confinement. Masses of W and Z bosons are also determined by their interactions with the vacuum. Many other examples can be given. One of them, the vacuum polarization, is discussed in this chapter. It appears, that the vacuum is polarized by external fields (electromagnetic or gravitational) like an ordinary dielectric. External fields interact with virtual particles, and thus change vacuum properties. These changes are observable and in particular in the cosmological case may lead to inflation.

7.1 Vacuum polarization in electrodynamics

Let us begin with the simpler, better known, and what is most important, experimentally verified example of quantum electrodynamics. The vacuum, like a dielectric, is polarized by an external electric field, for example by a point-like charge Z. This field can be measured by scattering of charged test particles (figure 7.1).

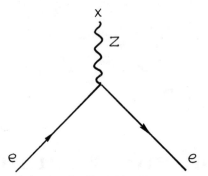

Figure 7.1: Scattering of electron in the field of heavy charge Z.

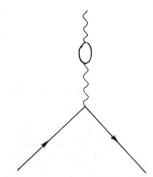

Figure 7.2: Charge screening by virtual e^+e^- pairs.

As it was already mentioned, vacuum fields are altered in the presence of an external charge Z. Virtual electrons tend to be closer to this charge, while positrons tend to be away from it (if $Z > 0$). The corresponding diagram is shown in figure 7.2. This phenomenon leads to the screening of the electric charge in complete analogy with the case of a dielectric. The charge at large distances proves to be smaller than that at small distances.

One usually considers the dependence of charge on momentum transfer and not on distance. There is an evident connection between these two ways of presentation. Distance and momentum are Fourier conjugate variables, and thus the bigger is the momentum transfer, the smaller is the distance, and vice versa. It can be shown that charge squared depends on the momentum transfer q in

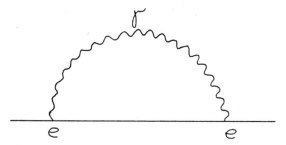

Figure 7.3: Emission and absorption of a virtual photon by electron.

the limit of large q^2 as:

$$Z^2(q^2) = \frac{Z_0^2}{[1 - \alpha\, b \ln(q^2/q_0^2)/2\pi]} \qquad (7.1)$$

where $\alpha = 1/137$ is the fine structure constant, and b is a constant depending on a number of particle species contributing into vacuum polarization (this is the number of different loops which can be drawn on the diagram of fig. 7.2). This formula reflects the fact that charge is bigger at smaller distances. A particle is no longer point-like, but it acquires some spatial charge distribution, thanks to the interaction with vacuum.

This leads to the well known ultraviolet instability of quantum electrodynamics. The point is that if at vanishingly small distance the charge is finite, than the charge at large distances (the Coulomb charge) must be zero. This indicates that the theory is logically inconsistent. However, it has been established in recent years that apart from the electromagnetic interaction there are some other interactions, which lead to modification of charge behaviour at small distances and to charge antiscreening.

The effects of vacuum polarization can be observed experimentally. It is measured with very high precision in atomic physics. It is known that the levels $2S_{1/2}$ and $2P_{1/2}$ of hydrogen atom are degenerate even if relativistic corrections are taken into account. In quantum field theory this degeneracy is, however, destroyed. The splitting of the levels is caused by two effects. The first is that the electron can emit and adsorb a virtual photon (figure 7.3). As a result the electron interaction with the Coulomb field of a nucleus

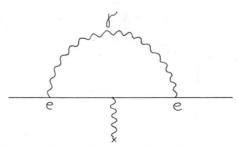

Figure 7.4: Modification of electron interaction with external field by emission and absorption of a virtual photon.

is altered (see figure 7.4). Besides that the emission of a virtual photon acts as though it increases the mean value of the electron momentum. Consequently the electron wave function becomes less localized and its value in the neighborhood of the nucleus decreases. The level $2S_{1/2}$ is very much affected by that. It goes up by 1000 MHz while level $2P_{1/2}$ is almost not shifted, since its wave function vanishes at the centre. The position of an energy level is most sensitive to the behaviour of the wave function near the nucleus where the electric field is the strongest.

The second effect is due to modifications of the Coulomb law by vacuum polarization (figure 7.1). This shifts the $2S_{1/2}$ level by about 25 MHz and for the above mentioned reasons has only a very small impact on the $2P_{1/2}$ level (the latter goes down by about 17 MHz due to both considered effects).

The level splitting is called the Lamb shift, and was experimentally confirmed with very high precision not only in the hydrogen atom, but also in the muonic atoms (an atom in which a negatively charged muon takes the place of an electron). The latter is especially interesting because its Bohr radius is $m_\mu/m_e = 210$ times smaller than that of the ordinary hydrogen atom, and vacuum polarization effects are stronger. In the ordinary hydrogen atom vacuum polarization gives about 3 percent of the total energy level shift, whereas it dominates in muonic hydrogen, where it shifts the level downwards by about $0.2\text{eV} = 3 \times 10^8$ MHz.

Experimental investigation of the splitting between $2S_{1/2}$ and $2P_{1/2}$-levels is especially sensitive to specific effects of quantum field theory since the splitting is absent in ordinary quantum mechanics. Such experiments are called null experiments, because there the small effect is not concealed by a big signal. Thanks to laser technology, we can now measure precisely the frequency of $2P \rightarrow 1P$ or $3P \rightarrow 1S$ lines and find in this way the shift of the ground state 1S caused by virtual particles. It is 8 times bigger than the corresponding shift of the 2S level in accordance with the value of $|\Psi(0)|^2$.

Vacuum polarization corrections to the Coulomb law are very simple in the extreme cases of small ($r \leq 1/m$) distances:

$$\phi(r) = \frac{e}{r}\left[1 - \frac{2\alpha}{3\pi}\left(\ln(m_e r) + C + \frac{5}{6}\right)\right] \qquad (7.2)$$

and large distances:

$$\phi(r) = \frac{e}{r}\left[1 + \frac{\alpha}{4\sqrt{\pi}}(m_e r)^{-3/2}\exp(-2\,m_e r)\right] \qquad (7.3)$$

where $C = 0.577$ is the Euler's constant. We shall not present here the quantum field theory calculations needed to obtain this result but their physical grounds are very simple.

Let us note that the inverse square law is only approximate according to expression (7.3), but the deviations from it for $r \geq 1/m = 10^{-11}$cm are extremely small.

If particles constantly appear and disappear in vacuum one could try to pull them out of there and make them real. Indeed this is possible. For example a sufficiently strong electric field can create pairs of charged particles. The probability of such a process can be estimated as follows. Let $W(l)$ be probability that a virtual electron and a positron are born at the distance l. Expression (7.3) indicates that $W(l) \approx \exp(-2m_e l)$. If l is big enough i.e. if the difference of potentials of e^+ and e^- in electric field is bigger than the sum of their masses:

$$e(\phi_1 - \phi_2) = eEl \geq 2m$$

then the creation of e^+ and e^- pairs is energetically allowed. The creation probability can be estimated to be $W \approx \exp(-const \times$

m^2/eE). This simple argument, of course, does not allow us to find the numerical coefficient under the exp sign. The exact formula for small field E is:

$$W = (\alpha E^2/\pi^2)\exp(-\pi E_0/E) \qquad (7.4)$$

where $\alpha = e^2 = 1/137$ is the fine structure constant, $E_0 = m^2/e = 1.3 \times 10^{16}$ V/m is the critical field. This expression can be obtained by considering the tunnelling transition of an e^+e^- pair from the classically forbidden region to the allowed one. The physical meaning of the exponential factor in (7.4) is the same as in the case of a particle passing through a potential barrier, which is higher than the energy of the particle.

An electric field of the order of E_0 is the absolute upper bound on possible traditional type accelerators since stronger fields would be screened by the created particles. It might be interesting that the size of an accelerator needed to produce intermediate bosons of Grand unified theories with mass $m = 10^{15}$ GeV, would be about 1000 km even if a critical field were used.

Maybe a more effective way to accelerate particles would be one using a laser beam, and not, as it is done now, static (or slowly changing) fields. Static fields are maximal on their borders, while a laser beam can be focused along the path of accelerated particles.

There is a paradox connected with particle creation. The rate of energy density change of created particles $mc^2\frac{dn}{dt}$ should be approximately equal to the work done by the field on the particles:

$$mc^2 \frac{dn}{dt} = jE$$

where $j = env \leq enc$ is the current density. Therefore:

$$\dot{n} \leq enE/mc$$

or

$$\ln(n/n_0) \leq e\int E\frac{dt}{mc}$$

Thus, if there were initially no particles then no particles should appear. This paradox was formulated by S.Hawking in connection with a gravitational field. The reason for the vanishing of particle production, is that their creation rate is proportional to their density $\dot{n} \sim n$. If the initial condition is $n = 0$, then the solution must be $n(t) \equiv 0$. This paradox was solved by Ya.B.Zeldovich and L.P.Pitajevsky. They noticed, that the particles are not created from nothing, but they exist virtually in vacuum. The electric field polarizes them, and the polarization is proportional to the field. As a result we come to the equation $\dot{n} \sim \sqrt{n}$ which has non-zero solution with zero initial condition.

This can be illustrated with a more simple but absolutely adequate example of the polarization of a hydrogen atom by an external electric field. The hydrogen atom is known to have zero electrical dipole moment:

$$d = \int \Psi^* \Psi x dx$$

But if the atom is polarized by external electric field, the $1S$ state is mixed with the $2P$ state:

$$\Psi = \Psi_{1S} + \delta \Psi_{2P}$$

and there appears a dipole moment $d \approx \delta$. Let us now calculate the rate at which the $2P$ state is produced. The probability that the atom is in the $2P$ state is:

$$W \sim \delta^2$$

and the correction to the energy is also proportional to δ^2:

$$E = E_{1S} + \delta^2 E_{2P}$$

But the Hamiltonian of the interaction is proportional to δ:

$$H_{int} \sim \delta E$$

Hence the probability of the creation of an admixture of $2P$ state is described by the equation:

$$\dot{W} = const\sqrt{W}$$

which has the nonvanishing solution $W \sim t^2$ with zero initial condition $W(0) = 0$.

7.2 Vacuum polarization by curved spacetime

The gravitational field polarizes vacuum like the electromagnetic one but since the gravitational field equations are nonlinear new interesting effects can arise. The classical Einstein equations have the form:

$$G_{\mu\nu} = \kappa T_{\mu\nu} \tag{7.5}$$

where $G_{\mu\nu} = R_{\mu\nu} - g_{\mu\nu}R/2$ (see (3.21)). Quantum corrections alter the energy momentum tensor of matter. It was already mentioned, that quantum corrections change particle masses and interaction constants. Moreover, the quantum energy of zero oscillations leads to the appearance of a cosmological constant i.e. to a contribution to the energy momentum tensor of the form $T_{\mu\nu}^{(1)} = \rho_{vac}g_{\mu\nu}$ (chapter 5). There are other quantum corrections due to change of zero oscillations by the gravitational field (vacuum polarization by gravitational field). These corrections vanish when the spacetime curvature approaches zero. Corrections to the energy tensor due to this effect were discussed by Ginzburg, Kirzhnits and Ljuboshin (1971).

The correction of the simplest form is:

$$T_{\mu\nu}^{(2)} = AG_{\mu\nu} \tag{7.6}$$

where A is a constant. This is implied by the tensor structure of $T_{\mu\nu}$ and by the covariant conservation law:

$$T_{\mu;\nu}^{\nu} = 0$$

This correction leads to quantum renormalization of the gravitational constant and is nonobservable.

A nontrivial gravitational vacuum polarization correction to $T_{\mu\nu}$ is:

$$T_{\mu\nu}^{(3)} = B_1 \left(2R_{;\mu\nu} - 2g_{\mu\nu}R - \frac{1}{2}g_{\mu\nu}R^2 + 2RR_{\mu\nu} \right)$$

$$\tag{7.7}$$

$$+B_2 \left(2R_{\mu;\nu\alpha}^\alpha - \partial^2 R_{\mu\nu} - \frac{1}{2}g_{\mu\nu}\,\partial^2 R + 2R_\mu^\alpha R_{\alpha\nu} - \frac{1}{2}g_{\mu\nu}R^{\alpha\beta}R_{\alpha\beta} \right)$$

The dimensionless coefficients B_1 and B_2 are not determined by the theory they are arbitrary renormalization constants. Both terms in equation (7.7) are covariantly conserved. Corrections (7.6) and (7.7) are local, i.e. they depend on spacetime properties at one point. Some nonlocal terms are also possible but their form in the general case is unknown.

The vacuum polarization corrections to the energy-momentum tensor in a curved spacetime are especially simple in the de Sitter metric. The high degree of symmetry , described in section 2 of chapter 5, allows one to determine the tensor structure of geometrical objects. In particular:

$$R_{\alpha\beta\gamma\delta} = H^2(g_{\alpha\gamma}g_{\beta\delta} - g_{\alpha\delta}g_{\beta\gamma})$$

$$\tag{7.8}$$

$$R_{\alpha\beta} = 3H^2 g_{\alpha\beta}$$

and finally:

$$\Delta T_{\alpha\beta} = const\, H^4 g_{\alpha\beta} \tag{7.9}$$

New solutions of the General Relativity equations appear when the quantum correction (7.7) is taken into account. Apart from the trivial solution corresponding to flat spacetime with $R_{\mu\nu\alpha\beta} = 0$ there exist solutions with nonvanishing $R_{\mu\nu\alpha\beta}$. In other words spacetime can be curved, and expanding even in the absence of real particles. These solutions are self-consistent in the sense that the spacetime curvature is caused by vacuum polarization and vice versa.

Let us demonstrate this in the case of de Sitter spacetime. Equation (7.5) without matter but with the correction (7.9) is:

$$H^2 g_{\mu\nu} = AH^4 g_{\mu\nu} m_{Pl}^{-2} \tag{7.10}$$

One should note that the hypothesis about the form of the metric is self consistent. Corrections to $T_{\mu\nu}$ in the de Sitter metric were inserted in the general relativity equations, and the solution was once again the de Sitter metric. Equation (7.10) has not only the trivial solution $H = 0$, but also $H = m_{Pl}A^{-1/2}$. It can be easily seen that A is proportional to the number of particles species that contribute to vacuum polarization.

Thus we have arrived at a model, in which exponential expansion is caused by quantum corrections to the energy momentum tensor. The possibility of such a regime was first noted by Gurovich and Starobinsky in 1979. Later Starobinsky (1980) proposed the inflationary model based on vacuum polarization effects. It can be checked that equation (7.5) with an energy momentum tensor given by (7.7) has the de Sitter solution (among many others). This solution is unstable and small deviations grow with time. This ultimately leads to transition from the exponential expansion to the Friedman one. The rate of growth of these perturbations depends on the above mentioned number of particles species. When N is big i.e. $N \approx 10^3$ then horizon, flatness and other problems can be solved. It was recently shown that N probably does not have to be that high if other corrections to $T_{\mu\nu}$ are considered.

Chapter 8

Baryosynthesis in the universe

Astronomical observations show that there is practically no anti-matter (i.e antiprotons, antineutrons, positrons) in the Universe. We are absolutely sure that our Galaxy consists only of matter because otherwise gigantic bursts of energy due to mutual annihilation of matter and anti-matter would be observed. Analogously it can be concluded that every observed galaxy contains one type of matter only (i.e. either matter or anti-matter). Colliding galaxies and galaxies immersed in the same cloud of intergalactic gas are observed and there is no trace of annihilation. So it is reasonable to assume that there is no antimatter in the Universe and thus we encounter the problem of charge asymmetry of the Universe. It is sometimes called baryon asymmetry, since there are baryons and no antibaryons. A small number of antiprotons is observed in cosmic rays, but it is only about 10^{-4} that of protons. It is interesting that the number of relic neutrinos must be almost the same as that of antineutrinos, so a large charge asymmetry at the modern time exists for strongly interacting particles only.

The phenomenon of baryon asymmetry of the Universe is especially weird because the number of baryons is much smaller than the number of relic photons:

$$N_B/N_\gamma = 10^{-9} - 10^{-10} \qquad (8.1)$$

It means that the numbers of protons and antiprotons were almost equal in the hot equilibrium primordial plasma at the temperatures $T \geq m_p$:

$$(N_p - N_{\bar{p}})/N_p = 10^{-9} - 10^{-10}$$

More precisely, one must speak at these temperatures about quarks and antiquarks . In any case there existed a very small excess of quarks over antiquarks. The existence of all the visible matter in the Universe is due to this small excess of particles over antiparticles.

Even 15-20 years ago the origin of this asymmetry was mysterious. The asymmetry was taken as an initial condition that could not be explained by physical laws. Models with initially cool baryon matter with subsequent heating leading to pair creation were discussed. Now it has changed. The asymmetry can be explained and its value estimated.

8.1   Baryon non-conservation and proton decay

The crucial point is the assumption that baryon number is not conserved. We shall now discuss this in detail. The most common baryons are protons and neutrons, which make up all the atomic nuclei. Apart from them there are many short lived baryons such as strange (hyperons), charmed etc. All these particles are members of one family united by the property that their number is conserved in all observed so far processes. A decaying neutron, for example, produces another baryon, the proton:

$$n \rightarrow p + e^- + \nu$$

Creation of baryons always proceeds together with creation of equal number of antibaryons, for example:

$$\pi^+ + p \rightarrow p + p + \bar{p} + \pi^+$$

Baryonic charge B, equal to (+1) for baryons and (-1) for antibaryons, was introduced in order to describe this phenomenon. The law of baryon conservation can be formulated as a baryonic charge conservation law in complete analogy with conservation of electric charge.

The stability of nuclei and of the proton which is the lightest nucleus, is the most impressive proof of baryon conservation. The proton lifetime has been measured to be longer than 10^{31} years.

There is, however, a very important difference between electric and baryonic charges. Electric charge is the source of a massless vector field (the electromagnetic field). The theory demands that the source of a massless field be conserved. No massless field and corresponding long range forces connected with baryonic charge have been found despite extensive searches (see section 4 chapter 3). Thus there are no theoretical grounds for baryon conservation. Moreover, grand unification theories, unifying electroweak and strong forces, predict that baryonic charge should not be conserved. As it has been already mentioned there exist vector fields X and Y in these theories that interact with quarks and leptons as follows:

$$X \leftrightarrow 2q, \quad X \leftrightarrow \bar{l}\bar{q}$$

and analogously for Y.

The electrical charge of X field quanta is $\pm 4/3$ and that of Y is $\pm 1/3$.

It can be easily seen that baryonic charge is not conserved in the chain of reactions:

$$q + q \leftrightarrow X \leftrightarrow \bar{q} + \bar{l}$$

since $B(q) = 1/3$, $B(\bar{q}) = -1/3$ and $B(\bar{l}) = 0$. Let us note that ΔB is integer in this process, $\Delta B = -1$.

A question immediately arises: why is the proton stable or, to be more exact very long-lived, if B is not conserved? This problem can be solved if the mass of X and Y bosons is high enough. The simplest unification models based on $SU(5)$ symmetry, predict that

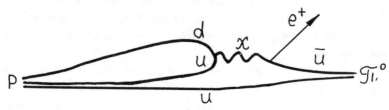

Figure 8.1: Proton decay into π^0 and e^+.

$$P \left\{ \begin{array}{l} \underline{d} \\ \underline{u} \\ \underline{u} \end{array} \right.$$

Figure 8.2: Quark content of proton.

$m_X \approx m_Y \approx 3 \times 10^{14}$ GeV (unfortunately as we shall see this appears to be not large enough).

The proton decays according to the diagram of figure 8.1. The pair $q\bar{q}$ forms π or K mesons and $\bar{l}$ can be a positron or a muon. Thus, we obtain reactions like $p \to \pi^0 e^+$, $K^0 \mu^+$ etc.

From a quantum mechanical point of view proton decay can be described exactly like neutron decay . The proton is a bound system of three quarks (see figure 8.2). There exists an interaction that transforms two quarks into a superheavy boson X. The amplitude of this transition C_1 is proportional to:

$$C_1 \approx em_p/(m_X + m_u - m_p)$$

Thus a state is created which is described by wave function $\psi_p + C_1\psi_{Xu}$ shown by the central part of the diagram in figure 8.3. Since the creation of an X boson is a virtual process, the transition of X to real particles e^+ and $\bar{u}$ must be considered. The existence of such processes is predicted by the unification theory. The amplitude

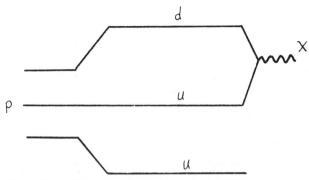

Figure 8.3: Transition of d and u quarks into virtual X boson.

of this transition is proportional to:

$$C_2 \approx e m_p / (m_X + m_u - m_e - m_\pi)$$

The difference between proton and neutron decay is that the X boson is much heavier than the W boson and that in the first case the mass defect is in order of magnitude to the proton mass. The fine structure constant $e^2/4\pi$, which equals $1/137$ in electrodynamics, is equal to about $1/40$ in the unification theory. This is caused by charge renormalization at high energies $E \approx 10^{14}$GeV.

If the X-boson mass were smaller than proton mass, it could be created in the process of proton decay, and the decay would be allowed in the first order of perturbation theory. Its probability would be proportional to C_1^2 but since $m_X > m_p$ the decay is possible only in the second order of perturbation theory. The probability of decay is thus proportional to:

$$\Gamma \approx e^4 (m_p/m_X)^4 m_p$$

and the proton lifetime $\tau \equiv \Gamma^{-1} \approx 10^{-20}(m_X/m_p)^4 s$. Since quarks do not exist as free particles they must form colorless hadrons like π or K mesons. This corresponds to the third part of the diagram of figure 8.4. The probability of proton decay into a definite channel is determined by the probability of transition of an X boson into a quark-lepton pair and by the probability of production of a specific hadron from the quark-antiquark pair. The uncertainties of the

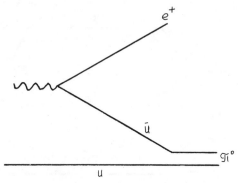

Figure 8.4: Transition of a virtual X-boson into e^+ and $\bar{u}$ and production of π^0 out of $u\bar{u}$.

estimate of Γ are connected with the poor knowledge of the quark wave function in a proton and the probabilities of quark fusion into hadrons. This, however, cannot change our estimate by more than an order of magnitude.

The probability of processes with baryon number non-conservation should grow as the energy of colliding particles grows. It is clear that the lifetime of a single proton does not depend on its energy (not taking into account relativistic time delay). However the process of proton extinction like e.g.:

$$p + p \to \pi^0 + e^+ + p$$

is possible. This process is described by the diagram of figure 8.5. The scattering cross section and so the probability of the process with baryonic charge non-conservation is proportional to square of the center of mass energy of the colliding particles. The relative probability of such processes is of the order of unity at $E \approx m_X \approx 10^{14} - 10^{15} \mathrm{GeV}$.

For a numerical estimate of the proton lifetime let us substitute $m_X = 3 \times 10^{14}$ GeV and get $\tau_p \approx 3 \times 10^{29}$ years. The result of exact calculations is a little bit higher but it is still below existing experimental bounds. This proves that the simplest model is not true but the idea of unification is by no means rejected because in a more complicated theory one can get a higher value of τ_p.

Some experimental groups have recently reported, that they ob-

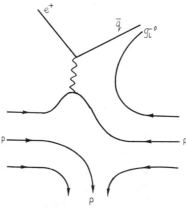

Figure 8.5: Baryon charge non-conservation in the reaction $p + p \rightarrow \pi^0 + e^+ + p$.

served something that could be interpreted as proton decay but this might have been produced by the background of cosmic rays. Now the most sound argument in favor of baryon non-conservation is the baryon asymmetry of the Universe. The development of experimental technique can improve the bound on τ_p up to 10^{34} years. If, however, even at this level of sensitivity proton decay is still not found, then cosmology presents the only remaining way to test the grand unification theories.

8.2 Nonstationarity of the universe and charge asymmetry

For explanation of the baryon asymmetry not only baryon non-conservation is necessary, but also the fact that the Universe is not stationary and that particles interact differently from antiparticles. In contrast to the hypothesis of baryon non-conservation both these facts are very well established. The idea that baryon asymmetry of the Universe was generated because of these phenomena was proposed 15 years ago by A.D.Sakharov. Later on a lot of concrete models of its realization were worked out.

There will be no excess of baryons over antibaryons, if baryonic charge is not conserved, but the interactions of particles and antiparticles are identical i.e. if charge symmetry or C invariance is exact. If the plasma is in a state of thermodynamical equilibrium, then the numbers of particles and antiparticles are the same independent of C invariance. This is a consequence of the CPT theorem, which states that physical processes do not change under simultaneous space reflection, time reversal, and particle-antiparticle conjugation.

CPT invariance is implied by general properties of the theory, like Lorentz invariance and analyticity. According to the CPT theorem masses, lifetimes, and statistical properties of particles and antiparticles, should be the same. This leads to equal numbers of particles and antiparticles in equilibrium plasma, since the distribution functions depend only on masses and statistical weights of particles. It should, however, be noted that if particles possess conserved non-zero charge then there will always be an excess of particles over antiparticles or vice versa, corresponding to the charge density in the plasma. The chemical potential is used for the description of distribution functions in this case.

We assume that baryonic charge is not conserved, so regardless of the initial conditions the primordial plasma must possess $B = 0$ if thermodynamic equilibrium is established. The Universe expands, it is nonstationary, therefore there can be a deviation from the equilibrium. This deviation is determined by the ratio between the rate of the expansion of the Universe $H = \dot{a}/a$, and the rate of the reactions that restore the equilibrium.

The expansion of the Universe does not disturb the equilibrium of massless particles. The cosmic background radiation is a good example. Indeed the interactions of 3K-photons between themselves and with matter are negligibly small, while the spectrum is nevertheless described by the equilibrium Planck form with the adiabatically decreasing temperature. This is true for any massless particles. To prove that let us consider kinetic equations in the expanding Universe. These are obtained by changing the ordinary derivatives to covariant ones. One obtains for the case of the metric (2.1):

$$\frac{\partial n_i}{\partial t} = H p_i \frac{\partial n}{\partial p_i} + S(n) \qquad (8.2)$$

where $n_i(p_i, t)$ is the density of particles i with the momentum p_i, H is the Hubble parameter and S is the collision integral which depends on densities of all the particles interacting with i.

It can be easily seen, that for massless particles equilibrium distribution functions satisfy these equations. Let us check this for the simple case of Boltzman statistics i.e. $n_i = n_{ieq} = \exp(-E_i/T)$. Fermi and Bose statistics, and the case of a non-zero chemical potential are left to the reader as an exercise. Calculating the derivative on the left-hand side of the equation we get:

$$\frac{\partial n_i}{\partial t} = \exp(-E/T)\frac{\dot{T}E}{T^2} = -\exp(-E/T)\frac{HE}{T} \qquad (8.3)$$

where the equation $\dot{T} = -HT$ is used. Since for massless particles $p = E$, and the collision integral vanishes for $n = n_{eq}$, the equilibrium distribution functions with time dependent temperature satisfy equation (8.2) in the nonstationary case. Thus for massless particles the equilibrium is always maintained despite the expansion of the Universe. In the case of massive particles the deviation from equilibrium is proportional to $(m/T)^2$ (if $m \ll T$). This deviation is large if $m > T$ but at such temperatures the concentration of massive particles is small due to the factor $\exp(-m/T)$ and so they cannot play an important role. Thus, the baryon asymmetry in this model was generated when the temperature of the primordial plasma was about the X boson mass i.e. $10^{15}\text{GeV} \simeq 10^{28}\text{K}$. In this scenario the baryonic charge evolved as follows. At the initial moment when T was probably equal to $T_{Pl} = 10^{32}\text{K}$ the value of B was arbitrary, and depended only on the mechanism of the "creation of the Universe". As the plasma reached a state of thermodynamical equilibrium the value of baryonic charge should disappear with an accuracy $\approx (m_X/T)^2$. This is already a lot, since the observed ratio of baryons to photons is $N_B/N_\gamma = 10^{-9} - 10^{-10}$ see eq.(8.1) and the ratio $(m_X/T)^2 > 10^{-8}$. However the inflation led to an exponentially small density of baryonic charge if there were no scalar condensates with nonzero and nonconserved baryonic charge. It should also be taken into account that at high temperatures X bosons are massless and they acquire mass at a temperature of order m_X. This is connected with the restoration of symmetry at high temperatures, indicated by Kirzhnits and Linde in 1972. Vector bosons acquire mass due to the above mentioned Higgs mechanism. It resembles symmetry restoration when a ferromagnet is heated. At low temperatures a ferromagnet can spontaneously magnetize. There appears a magnetic moment and because of its particular direction the symmetry with respect to rotations is broken. When the ferromagnet is heated above the Curie temperature, the magnetic moment is destroyed by the thermal energy, and the symmetry is restored. The same happens in field theory where the Higgs field condensate $\langle \phi \rangle$ takes the place of the magnetic moment (order parameter). When the temperature is small enough, the state that is energetically most favorable is that with condensate $\langle \phi \rangle \neq 0$. This results in a definite direction in the internal space, determined by the direction of $\langle \phi \rangle$ and thus the symmetry is spontaneously broken and the vector bosons acquire mass proportional to $\langle \phi \rangle$.

Therefore even if an initial charge asymmetry was present it should have disappeared during the epoch when $\langle \phi \rangle = 0$. It could appear once more only below the Curie temperature of the Higgs field when X bosons acquire nonzero mass and their distribution function becomes nonequilibrium. The degree of deviation from thermal equilibrium is determined by the ratio between the expansion rate $H = \dot{a}/a$ and the rate of particle reactions Γ. Accordingly the baryon asymmetry should be proportional to the ratio H/Γ when the temperature of the primordial plasma is $T \approx m_X$. In a simple model this ratio is about $0.1 - 0.01$.

8.3 Baryosynthesis and charge conservation

Before we proceed further let us note, that the standard $SU(5)$ model predicts non-conservation of both baryon B and lepton L numbers. Their difference however is strictly conserved in this model. Thus the value of $B - L$ is determined by the initial conditions.

In the case of quantum creation of the Universe (or as one says "creation from nothing") the resulting state should have all the quantum numbers of vacuum: zero energy, vanishing electric, and other conserved charges. If $(B - L)$ is conserved then its initial value should also be zero. It must be noted that the conservation of $B - L$ is not obligatory and there are a lot of models that do not predict it. Initial values of B and L are then arbitrary but one can expect that they are small. It is generally believed, that the Universe was created due to the gravitational interaction which most probably conserves B as well as L. Therefore the initial values of B and L should be equal to those in the initial vacuum state.

However there is no reliable way to calculate the initial values of B and L and strictly speaking, they can be arbitrarily big. Here once more inflation may help us. Regardless of the initial values of B and L they went down during the time t_i of inflation by $\exp(Ht_i) \geq 10^{30}$ and so they can be neglected. The baryosynthesis leading to the excess of baryons over antibaryons must take place after inflation.

This can be reinterpreted in the following way. If a noticeable baryonic charge is left after inflation then there was a small admixture of ordinary matter, $10^{-9}\rho_{vac}$, to the scalar field condensate energy density. Going backwards in time one can see that this small admixture quickly starts to dominate in the energy momentum tensor. The density of matter with the equation of state $p = -\rho_{vac}$ is constant, while the ordinary matter density grows quickly with the redshift, $\rho \approx (1 + z)^4$. The redshift increases exponentially $1 + z \approx e^{Ht}$. Thus already after $\Delta t = 7H^{-1}$ the ordinary matter energy density is 10^3 times larger than the scalar field energy density. Inflation is then impossible. This contradiction indicates that baryosynthesis took place after inflation.

8.4   Breaking of charge invariance

Before discussing a concrete model of baryosynthesis let us first consider breaking of C invariance i.e. the difference between properties of particles and antiparticles. As it was already mentioned the CPT-theorem is valid in quantum field theory. However, there is no reason for a theory to be invariant with respect to separate C, P, and T, transformations. It was proved by experiment that P-symmetry is broken in weak interactions. This is the well known phenomenon of parity non-conservation. After the discovery of parity non-conservation it seemed first that physical laws are invariant with respect to simultaneous P and C transformations. It has been found later that even this symmetry is only approximate. The breaking of both C and CP allows the mechanism of baryosynthesis discussed in the following section.

CPT-invariance states that masses of particles and of antiparticles are equal. The lifetimes of nonstable particles and their corresponding anti-particles should also be equal. If there are however several decay channels then because of CP violation the probabilities of charge conjugated processes may not be equal [1]. In the case of X and Y bosons which we are interested in the probabilities of the following decays:

$$\Gamma(X \to qq) - \Gamma(\overline{X} \to \overline{q}\overline{q}) = \Delta\Gamma$$

$$\Gamma(X \to \overline{l}q) - \Gamma(\overline{X} \to ql) = -\Delta\Gamma$$

can be different, but $\Gamma(X \to qq) + \Gamma(X \to \overline{l}q) = \Gamma(\overline{X} \to \overline{q}\overline{q}) + \Gamma(\overline{X} \to ql)$. Since $\Delta\Gamma \neq 0$ a different number of quarks and antiquarks are produced in the decay of equal amount of X and Y. Thus the baryonic charge density becomes nonzero.

The value of $\Delta\Gamma$ is small since it appears at least in the second order of perturbation theory. There is an additional suppression in the simple $SU(5)$ model connected with its high symmetry. Thus

---

[1] A similar phenomenon has been observed in experiment. The probabilities of K_L^0 meson decay into the channel $K_L^0 \to \pi^- \mu^+ \nu$ and into the charge conjugate one $K_L^0 \to \pi^+ \mu^- \overline{\nu}$ are different and their ratio is 0.997.

the value of N_B/N_γ predicted by the theory proves to be too low. So we see that the simplest version of the SU(5) Grand Unification Model contradicts experiment not only in its prediction for proton decay but also for the baryon asymmetry of the Universe. These discrepancies can be avoided in more complicated models with some additional particles.

8.5 Baryon asymmetry in grand unification theories and topological baryon non-conservation

Baryon asymmetry in Grand Unification Theories appears as follows. X bosons go out of equilibrium at $T \approx m_X \approx 10^{14} - 10^{15}$GeV. The deviation is of the order of H/Γ where $\Gamma \approx \alpha m_X \approx m_X/40$ is the decay width of the X-boson. It is the inverse of its lifetime. The ratio H/Γ is estimated to be approximately 3×10^{-2} which leads to an additional suppression of baryon production apart from the smallness of C and CP violating effects. Because of a small nonequilibrium of X-bosons, baryoproduction in their decays is not compensated by other processes. Hence there appears an excess of quarks over antiquarks. At smaller temperatures, when B is practically conserved, quarks and antiquarks annihilate except for that small excess that survives.

In this model the baryon asymmetry is generated at high temperature $T \approx 10^{14}$ GeV and during a time of about 10^{-34}s. Recently some models were formulated in which the asymmetry arises much later at "small" temperatures $T \approx 10^{3}$ GeV. In these models the baryon non-conservation is large at $T = T_0 \simeq 10^{3}$ GeV but at smaller temperatures or energies it is exponentially suppressed $W \approx exp(-T_0/E)$. The analogous suppression in grand unification models is by a power law. Baryosynthesis at small temperatures is connected with topologically nontrivial solutions in field theory, or with condensates of supersymmetric particles. A more detailed discussion of these models is however beyond the scope of our book.

We shall only note, that quantum corrections may lead to non-conservation of currents which are classically conserved. This is called a quantum anomaly. In particular, because of this anomaly, the baryonic current of left-handed quarks is not conserved (these quarks participate in interactions with W bosons). The anomalous term is a total derivative in the effective Lagrangian and so it should be unobservable. There are however field configurations in the quantum theory of nonabelian gauge fields which are solutions of equations of motion and in which the potentials do not vanish at infinity, but the field strength does. Such configurations of field cannot be removed by a nonsingular gauge transformation since they have a nontrivial topological structure. They cannot be obtained by a continuous, deformation of a gauge transformation which is close to unity. Such solutions are called instantons [37]. Because of them anomalous baryonic charge non-conservation arises. This non-conservation was first noticed by t'Hooft. However it is exponentially small, it is of the order of magnitude of $\exp(-1/\alpha)$. The effect becomes stronger in the presence of magnetic monopoles (Rubakov 1981, Callan 1982). Roughly speaking, the proton decay probability becomes of the order of unity in the vicinity of a magnetic monopole. The monopole acts as a catalyst of the proton decay process. It has, however, little impact on the baryon asymmetry of the Universe, since the number density of magnetic monopoles should be very small. It might happen that some classical objects, similar to magnetic monopoles, appearing in electroweak interaction theory, will produce considerable baryon asymmetry (Kuzmin, Rubakov, Shaposhnikov 1985). The question is whether such objects can be abundantly produced at a temperature $T \approx T_0 \approx 10^3$ GeV.

Let us note that the baryon asymmetry produced in inflationary cosmology can be $10 - 100$ times bigger than in the standard Friedman model . As it was mentioned above the asymmetry is suppressed by the small deviation of X bosons from thermodynamical equilibrium. In the inflationary model, however, the initial state can be highly nonequilibrium. As we already know, particle number density up till the end of inflation is practically zero and all the energy is contained in coherent oscillations of the scalar field. This scalar field produces all other particles, and among them the superheavy bosons of grand unification. Their density is far from equilibrium and so baryon asymmetry is not thermodynamically suppressed (Dolgov, Linde 1981).

8.6 Baryon asymmetry and black holes

Here we consider another model of baryosynthesis which operates even if baryonic charge is conserved (Hawking 1974, Zeldovich 1976, Dolgov 1980). Its mechanism is based on quantum evaporation of black holes. If interactions between created particles are neglected, the number of particles and antiparticles emitted by a black hole should be the same. This is not so if the interactions and C invariance breaking are taken into account. Let us consider a simple example. Let heavy unstable particles A be vaporized by a black hole and the probabilities of two different B conserving decays $A \rightarrow H + \overline{L}$ and $A \rightarrow \overline{H} + L$ be different. Let $H, (\overline{H})$ be a heavy baryon and $L, (\overline{L})$ be a light baryon (and corresponding antibaryon). The probability of backcapture by the black hole of a heavy baryon H, is larger than that of a light baryon L. Thus, baryonic charge is accumulated in the external space while antibaryons are conserved in the disappearing black hole. So gravitation leads to macroscopic baryon non-conservation, while baryonic charge is conserved microscopically. With a suitable choice of parameters, the result $N_B/N_\gamma = 10^{-9} - 10^{-10}$ can be obtained. Models in which baryonic charge is not conserved are, however, more attractive and only the absence of proton decay makes one suspect that they might not be valid.

The model discussed here is similar to the one described at the end of the previous section. Both models are based on classical, topologically nontrivial solutions (they are monopoles and instantons in the first case and black holes in the second). Baryonic charge is conserved in both models in the frameworks of perturbation theory, and non-conservation is connected with nonperturbative effects (the expression $\exp(-1/\alpha)$ cannot be obtained in perturbation theory, not even by summing up the perturbative series expansion).

Let us note how our views on proton stability have changed

during the last 20 years. It was usually said before that "our existence proves that the proton is stable" and now just the opposite statement is taken to be true "our existence proves that the proton is not stable". In the last case we mean that if baryonic charge is conserved then a Universe suitable for life is not created. Note in conclusion that the probability of baryonic charge non-conservation should be significant at high energies but sufficiently small at "room" temperature and low energies.

Chapter 9

Quantum creation of the universe

9.1 Creation from nothing and the energy conservation law

The two basic problems of this chapter are the following: what came before the very early Universe $t < t_{Pl} = 10^{-43}s$, and how was the Universe created ? Classical cosmology leads to singularities when $t \to 0$ i.e. to infinite values of energy density, pressure and to vanishing scale factor and comoving volume. Classical cosmology is not complete since there is no theory describing the earlier Universe. The initial state of the Universe is usually called a "singularity" but none knows what really happened there. Surely at $t < t_{Pl}$ the classical theory of gravity is inapplicable and probably quantum gravitational theory, not yet formulated, could solve the above mentioned problems.

Earlier there was a model of a pulsating Universe according to which the expansion of the Universe alternates with contraction. This scenario, however, did not settle the problem of the creation of the Universe. It only shifted it several cosmic cycles back. The point is that the entropy of the Universe should increase after each cycle and so the maximum radius in each consecutive cycle must be bigger. Therefore there cannot be infinitely many cycles unless it is assumed that entropy decreases when the Universe contracts or that physical laws are drastically changed at the Planck scale (M.A.Markov). The hypothesis that the entropy might decrease was also proposed by S.Hawking. This fantastic result is based on a cosmological model in which time is compact i.e time is defined not on an infinite line but on e.g. a circle but his idea was later criticized.

As quantum gravity theory was being developed there appeared to be a possibility to make at least qualitative statements about properties of the Universe at $t = t_{Pl}$ or even earlier. There appeared a modified model of a pulsating Universe in which the scale factor bounces near the singularity, because the Friedman regime $a = a_0 t^\alpha$ changes into the de Sitter $a(t) = a_0 \cosh(Ht)$, so that the scale factor was very small but nonzero at $t = 0$. Thus, all physical quantities remain finite. This model can be obtained from a quantum generalization of Einstein's gravitational theory. Additional terms non-linear in R :

$$\int d^4x \sqrt{-g} \left(a_1 R_{\mu\nu} R^{\mu\nu} + a_2 R^2 + a_3 \, \partial^2 R \right)$$

are added to the ordinary gravitational action:

$$S = \int d^4x \sqrt{-g} \, R$$

The coefficients a_1, a_2 and a_3 are assumed to be of the order of $l_{Pl}^2 = M_{Pl}^{-2}$. These are so-called radiative corrections to the gravitational action (see section 2 of chapter 7).

These corrections are analogous to vacuum polarization by electric charge. The Lamb-Rutherford effect discussed above, is caused by vacuum polarization in an atom, and is observed in experiment. Gravitational corrections are not proved by experiment but the experience with electrodynamics makes us believe they really exist.

It is essential that the equations with quantum corrections may lead to the de Sitter expansion law $a(t) = a_0 \cosh(Ht)$, which could help to avoid singularities. The gravitational vacuum polarization has been discussed in chapter 7 and the bouncing model will be presented later on in this chapter.

It seems that now the most popular is the idea that the Universe was created from nothing. It is assumed of course that the Universe was born without violating any known physical law, and that this process is described by quantum gravity and no energy for the creation of the Universe is needed. The probability of creation should be high enough.

First of all it must be stressed that such a process is not in contradiction with the energy conservation law. There are two aspects of the energy conservation law: global and local. The local energy conservation law can be written as:

$$dE = -pdV$$

for any local volume element dV. Each volume element has definite positive energy which can be changed by interactions with other volume elements. Special relativity has taught us that the energy of matter must not be zero. It is equal at least to the mass of a matter sample. General relativity has changed this point. It has been proved that the energy can decrease because of the gravitational mass defect. The idea that some matter can have zero energy has been realized only in the sixties - fifty years after the formulation of General Relativity. Thus there could exist a Universe with zero total energy. This would be a closed Universe. When such a Universe was born the most important physical law, the law of energy conservation would not be violated. The energy before and after the creation would be equal to zero. There would be no matter at first (and there would be no space as we understand it). Matter would appear after the Universe is created but the gravitational mass defect would be equal to the mass itself and so the total energy would remain zero.

Let us consider the global energy conservation law in General Relativity and the phenomenon of the gravitational mass defect. The mass defect exists not only for gravity, but also for any kind of interaction. It became known due to special relativity which established the equivalence of mass and energy. The mass of a bound state of two particles should be smaller than the sum of their masses, measured separately, since the interaction energy must be taken into account. This has been found experimentally, e.g. for atomic nuclei. The mass of a deuteron (bound state of proton and neutron) is smaller than the sum of the masses of the separate proton and neutron:

$$m_D = m_N + m_p - \Delta E$$

The nuclear mass defect $\Delta E = 2.2$ MeV is caused by the nuclear interaction between proton and neutron.

The same effect can be found in a double star system. Let the stellar masses be M_1 and M_2. The mass of the system is smaller than the sum of the masses and the difference is equal to the binding energy [1]:

$$M = M_1 + M_2 - \frac{GM_1M_2}{r}$$

Here GM_1M_2/r is the potential energy of two stars which causes the gravitational mass defect.

The gravitational mass defect is important for many observable properties of neutron stars. The mass of a star is calculated as follows:

$$M = \int \rho(r)4\pi r^2 dr$$

At first sight, there is no gravitational mass defect in this equation. The volume , however, in a strong gravitational field is different from that in the flat spacetime. The volume of a spherical sheet with thickness dr is the product of the surface of this sheet $4\pi r^2$ and its thickness dr i.e. it is equal to $4\pi r^2 dr$ in flat space. The surface does not change in a gravitational field but the thickness does:

$$dl = \frac{dr}{\sqrt{1 - r_g/r}}$$

so the volume element is:

[1]The kinetic energy should also be taken into account, $E_{1,2} = M_{1,2}/\sqrt{1 - v_{1,2}^2}$.

$$dV = \frac{4\pi r^2 dr}{\sqrt{1 - r_g/r}}$$

The mass of a neutron star expressed in terms of the physical parameters such as energy density and volume can be written as:

$$M = \int \rho(r)\sqrt{1 - r_g/r}\ dV$$

where r_g is the gravitational radius of the star. The factor $\sqrt{1 - r_g/r}$ determines the gravitational mass defect of the star. Thus, the mass of a neutron star is smaller than the sum of the masses of the particles it contains.

These examples show that the mass of a body is smaller than the sum of the masses of constituents. Let us consider now a sphere with constant density ρ. The mass of the sphere is:

$$M = \rho V - \int \rho \phi dV$$

The second term in this equation is the gravitational mass defect. The mass can be rewritten as a function of radius of the body as follows:

$$M = \frac{4\pi}{3}\rho r^3 - \frac{16}{15}\pi^2 G\rho^2 r^5$$

The first term is the sum of the masses of the particles in the body. It is proportional to the square of the density and to the cube of its size. The second term is the gravitational mass defect. It is proportional to the square of the density and to the fifth power of radius.

It follows from this expression that the mass can be made vanishingly small. Indeed as more matter is added to the body whilst keeping the density constant, the gravitational mass defect grows faster than the sum of the constituent masses and M becomes zero at a certain point. The radius is then:

$$r = \sqrt{\frac{5}{4\pi G\rho}}$$

This radius has been calculated in Newton's theory without taking into account variations of the metric when $M \to 0$. Therefore the coefficient cannot be exact, but by the order of magnitude the result is correct. Also correct is the conclusion that adding extra matter to the body one can ultimately get zero total mass.

Another possibility to obtain zero mass is to compress a piece of matter so that the product ρr^2 grows. In this way the mass can be made as close to zero as one wishes.

There is of course no contradiction with the energy conservation law. The decrease of mass must be accompanied by the radiation with energy equal to the mass defect.

9.2   Properties of half-closed universe

The examples which were discussed above lead to the conclusion that the total mass (or energy) of a closed Universe is equal to zero. It is interesting to consider how a closed Universe would form when some extra matter is added to an initial state. To this end let us look how the mass of a spherical body, as measured by an external observer, will be changed by the addition of matter. To describe the change of geometry general relativity effects must be taken into account.

Let us first recall some properties of a closed world and of a spherical body. The closed Friedman universe can be described by the metric:

$$ds^2 = a^2(\eta)(d\eta^2 - d\chi^2 - \sin^2 \chi d\Omega^2)$$

$$(9.1)$$

$$d\Omega^2 = \sin^2 \theta d\phi^2 + d\theta^2$$

The hyperspherical angle χ defines the distance from the center of the coordinate system. It changes between 0 and π. The space described by this metric is homogenous and isotropic. An arbitrary point can be chosen as the center of the coordinate frame.

The geometry corresponding to the metric (9.1) is the geometry of a three-dimensional spherical surface in flat four-dimensional space. Let us introduce coordinate v exactly like we did in the case of the de Sitter metric. The spherical surface is given by the equation:

$$v^2 + x^2 + y^2 + z^2 = a^2 \qquad (9.2)$$

The interval in the 4 dimensional space (v, x, y, z) is given by:

$$dl^2 = dv^2 + dx^2 + dy^2 + dz^2$$

and on the surface of the sphere it is:

$$dl^2 = a^2(d\chi^2 + \sin^2 \chi d\Omega^2)$$

When the angle χ lies between 0 and $\chi_0 < \pi$ the metric (9.1) describes only a part of the Friedman universe. If $\pi/2 < \chi_0 < \pi$ then such a space is called a half-closed Friedman space. This part of the Friedman space is no longer isotropic but is spherically symmetric and the center is $\chi = 0$.

The metric in the neighborhood of a spherical body is:

$$ds^2 = \left(1 - \frac{r_g}{r}\right)dt^2 - \left(1 - \frac{r_g}{r}\right)^{-1}dr^2 - r^2d\Omega^2 \qquad (9.3)$$

where r_g is the gravitational radius of the body:

$$r_g = 2Gm = 2G \int_0^{r_0} \rho(r)4\pi r^2 dr \qquad (9.4)$$

In this expression the mass of the body has been written in the form of a volume integral, in order for the metric to be continued inside the body . The metric inside is:

$$ds^2 = (1 - 8\pi G\rho^2 r^2/3)dt^2 - (1 - 8\pi G\rho^2 r^2/3)^{-1}dr^2 - r^2 d\Omega^2$$

if the density is constant. We shall discuss only the three dimensional part of the above metric, which determines the space geometry:

$$dl^2 = (1 - 8\pi G\rho^2 r^2/3)^{-1}dr^2 + r^2 d\Omega^2$$

We show now that the geometry of this three dimensional hypersurface is identical to the geometry of the closed Friedman space i.e. it is the section of closed space at maximum expansion time. Let us introduce parameter a with dimension of length by the formula:

$$a^2 = 3/(8\pi G\rho)$$

and express the radial coordinate through the hyperspherical angle as:

$$r = a\sin\chi$$

The metric (9.5) can now be written as:

$$dl^2 = a^2(d\chi^2 + \sin^2\chi d\Omega^2)$$

which resembles the metric of three-dimensional closed space. The only difference is that the angle χ changes between 0 and π in the closed space, while in our case it is smaller than π.

Let us return to the description of the three-dimensional space as a hypersurface in four-dimensional space. The hypersurface corresponding to the Schwartzschild geometry in empty space satisfies the equation:

$$1 + (dv/dr)^2 = (1 - r_g/r)^{-1}$$

which has the following solution:

$$v(r) = 2\sqrt{r_g(r - r_g)} \qquad (9.6)$$

Inside the matter the three dimensional space is a part of the spherical hypersurface:

$$v^2 + r^2 = a^2 \qquad (9.7)$$

The geometry of the whole space can be represented as a matching of the hypersurface (9.6) in the empty space and the hypersurface (9.7) inside the body.

The $v(r)$ dependance obtained in this way is shown in figure 9.1.

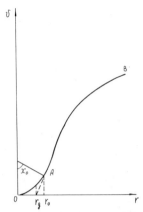

Figure 9.1: Gluing the Friedman and Schwartzschild geometries.

The first part OA corresponds to a part of the Friedman solution, obtained by rotation of OA around the Ov axis. Part AB corresponds to the Schwartzschild metric. This curve if continued down to the axis Or (dotted line) crosses it vertically at $r = r_g$. This continuation is not, however, done. Note that the smaller is the angle χ_0, the smaller is the curvature of the surface obtained by rotation of OAB. Vice versa: the bigger is the angle χ_0, the more the surface is curved, and the bigger is the curvature of the real space. When $\chi_0 = \pi/2$ the Friedman part forms half of the total closed space. Both curves OA and AB meet at the point A with vertical tangents. An external observer sees a black hole with gravitational radius a in the center of the coordinate system.

The surface corresponding to a half-closed space is shown in figure 9.2. The line OA corresponds to the Friedman part of the space. The ordinary Schwartzschild solution with mass equal to the mass of the half closed space is described by the line AB. Lines OA and AB' do not match because the derivatives dv/dr are different. The correct way to match the curves is $OACA'B$.

The mass measured by an external observer is expressed by equation (9.4):

$$m = \int \rho(r) 4\pi r^2 \, dr$$

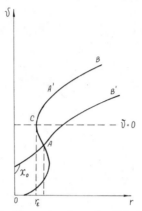

Figure 9.2: The same as in figure 9.1 but the Friedman space is more curved (half closed space).

Recall that $4\pi r^2 dr$ is not the volume element but that :

$$dV = 4\pi r^2 (1 - r_g/r)^{-1/2} dr$$

In the half-closed world coordinate r does not change monotonically when going from the center. It goes along the curve $OACA'B$. The latter describes a surface when rotated around axis Ov. A point with fixed r on this curve determines a circle with radius $r = a\sin\chi$ on the surface. In a three dimensional geometry it corresponds to the three dimensional sphere.

The picture, on which the points with $r = $ const are presented as circles, is shown in figure 9.3. Going from the central point $x = y = 0$ the radius grows until it reaches its maximum value at $\chi = \pi/2$ and then decreases back to r_0. r_0 is the gravitational radius measured by an external observer. Since $r_0 < a$ when $\chi_0 > \pi/2$ the external observer sees smaller mass then the sum of masses of separate elements of the body.

This can formally be proved by considering the formula expressing m through ρ:

$$m = \int \rho \, 4\pi r^2 dr$$

Since r does not change monotonically it is convenient to in-

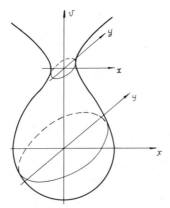

Figure 9.3: Almost closed Friedman space. Its mass can be arbitrarily low from the point of view of an external observer.

troduce a new variable which always increases. Let us choose the length along the curve OA in figure 9.2 as this variable. The mass in the new coordinates is:

$$m = \int \rho(q) 4\pi r^2(q) \frac{dr}{dq} dq$$

Adding matter (strictly speaking dust) to the inner part of the half-closed space i.e to the part of the sphere where $\chi > \chi_0$ makes the mass observed by the external observer smaller since $dr/dq < 0$ on the curve AC.

As the volume of the half-closed space increases (χ_0 increase with fixed a) because of addition of extra matter the mass decreases because the gravitational mass defect is larger than the mass added. Since the mass becomes smaller the gravitational radius is also smaller i.e. the surface bounding the half-closed space shrinks. Let us recall that there is the Schwartzschild singularity on this surface. Thus one can continuously obtain vanishing mass corresponding to the closed Universe.

Since the energy (or mass) of a closed Universe is equal to zero the transition from a universe with a small radius to the one with a big radius is possible. The most important point is that there is no violation of the energy conservation law.

9.3 Mass of the matter in the universe and its increase during inflation

It is naturally expected that the Universe was born with a volume of about l_{Pl}^3. The radius of curvature was also about l_{Pl}, all other parameters like mass (without gravitational mass defect), density etc. also had characteristic Planck values. Although the total mass of the newly-born closed Universe was zero, its positive part, that is the mass of matter without the gravitational mass defect, is about $m_{Pl} = 10^{-5}g$. We know on the other hand, that the mass of the matter in the Universe now, is at least $M \approx 10^{55}g \approx 10^{22}M_\odot$. It is interesting to ask where this mass has come from.

To this end let us consider the local energy conservation law:

$$dE = -pdV \qquad (9.8)$$

or in the equivalent tensor form:

$$T^\nu_{\mu,\nu} = 0$$

Let us take a small comoving volume dV. Because of its small size one can neglect its internal gravitational energy and take into account only the energy of matter inside the volume.

The differential form of the energy conservation law is:

$$\frac{d\rho}{dt} = -\mathrm{div}J \qquad (9.9)$$

where J is the current density describing the change of energy density inside the volume. The differential energy conservation law in General Relativity is:

$$T^\nu_{\mu;\nu} = 0. \qquad (9.10)$$

This single entity, i.e. the energy-momentum tensor, satisfies the energy conservation law instead of energy, current and momentum density considered as separate quantities. This tensor is the source of the gravitational field. The field equations impose the condition (9.10) resulting in particular in the energy conservation law.

Let us recall the Maxwell equations:

$$\text{div}\,H = 0; \quad \text{div}\,E = 4\pi q$$

$$\text{curl}\,E = -\partial H/\partial t; \quad \text{curl}\,H = \partial E/\partial t + 4\pi j$$

The first two equations are the constraints imposed on the fields E and H. The second two describe the connection between the fields and their sources. Because of the first two equations evolution of q and j is restricted by the charge conservation law. In the differential form this law is :

$$\partial q/\partial t + \text{div}\,j = 0$$

This law prevents from the production of electrical charge out of nothing. If one wants to create a charge somewhere in the space then one should produce a current to this point.

The same considerations prevent the production of mass out of nothing in nonrelativistic physics since the energy conservation law (9.9) has the same form as the charge conservation law.

In General Relativity the gravitational mass defect could completely eat the mass of matter up, so that the total mass of a system becomes zero. The amount of matter in the Universe may change arbitrarily without any contradiction with the energy momentum conservation law (9.10). It is essential for cosmology that during inflation this amount grows exponentially.

It is convenient to consider the variation of the amount of matter in the coordinate system in which $\sqrt{-g} = 1$. In a stationary gravitational field the equation expressing the dependence of energy density on the momentum flux is the same as in nonrelativistic physics (9.9). The coordinate system has been chosen so that gravity does not appear in this equation. The other conservation equations will contain terms expressing the work of the gravitational force $\nabla\phi$:

$$d(\rho v_j)/dt = \nabla_j p + \rho\nabla_j\phi$$

This work will be found explicitly in the case of a homogeneous and isotropic Universe with the Friedman metric:

$$ds^2 = dt^2 - a^2(t)(dr^2 + \sin^2 r d\Omega^2)$$

The energy momentum tensor can be written as:

$$T_\nu^\mu = (\rho + p)u^\mu u_\nu - p\delta_\nu^\mu$$

and the conservation equations are:

$$\frac{1}{\sqrt{-g}} \frac{\partial}{\partial x^\mu}(\sqrt{-g}T_\nu^\mu) = \frac{1}{2} \frac{\partial g_{\alpha\beta}}{\partial x^\nu} T^{\alpha\beta} \qquad (9.11)$$

There are no momentum fluxes, and stresses are uniform in the homogenous and isotropic Universe.

The left-hand side of equation (9.11) is the relativistic generalization of the conservation law (9.9) while the right-hand side represents the work done by gravitational forces. Since the metric does not depend on spatial coordinates one obtains:

$$\frac{1}{\sqrt{-g}} \frac{\partial(\sqrt{-g}T_0^\mu)}{\partial x^\mu} = \frac{1}{\sqrt{-g}} \frac{\partial(\sqrt{-g}\,\rho)}{\partial t} + \frac{\partial T_0^j}{\partial x^j}$$

Let us rewrite equation (9.11) in an integral form, substituting the explicit expressions for energy-momentum tensor and metric:

$$\frac{1}{a^3} \frac{\partial(a^3\rho)}{\partial t} + \operatorname{div} J = -3pH \qquad (9.12)$$

where $\dot{a}/a = H$ is the Hubble parameter. Let us consider an elementary volume formed by moving particles. This volume is expressed by Lagrange variables which are constant for each particle i.e they identify particles. A small comoving volume V is related to the Lagrange volume W by:

$$V = a^3 W$$

and W itself is equal to:

$$W = 4\pi\left(\frac{r}{2} - \frac{1}{4}\sin 2r\right).$$

In the limit of small r:

$$W \approx \frac{4\pi}{3} r^3.$$

The energy contained in volume V is:

$$E = \rho V = a^3 \rho W \tag{9.13}$$

and the rate of change of V with time:

$$\dot{V} = 3HV. \tag{9.14}$$

Let us multiply both sides of equation (9.12) by dW and integrate it over the Lagrange volume. The first term in the left hand side is equal to dE/dt according to (9.13). The second, according to Gauss theorem, is equal to $\int J ds$ and so it is zero since there are no currents across the borders of the volume. The right-hand side term can be connected with the change of physical volume according to (9.14). The equation of conservation can now be rewritten in the form $\dot{E} = -p\dot{V}$ which coincides with (9.8). The equation has now been proved in the framework of General Relativity.

Therefore, to ensure increase of energy $dE > 0$ during expansion $dV > 0$ the pressure must be negative. If the equation of state is $\rho = -p$ the energy density is constant and does not depend upon the volume variation dV. This was proved in chapter 5.

For this case equation (9.8) can be written in the form:

$$dE = \rho dV$$

so that E increases as volume V.

Hence, during inflation when $a(t) = \exp(Ht)$ and $p = -\rho$, the amount of matter in the universe grows exponentially fast as the volume of the Universe does.

9.4   Why does the universe expand ?

Let us discuss another important problem, namely, what was the origin of the isotropic and homogeneous expansion of matter which is called the Hubble expansion:

$$u = Hr.$$

Let us imagine a sphere with radius r in the expanding Universe. The energy conservation law states:

$$\frac{1}{2}\left(\frac{dr}{dt}\right)^2 = \frac{GM}{r} + const \qquad (9.15)$$

The equation of motion is obtained by differentiating the above equation with respect to time. In Newtonian physics mass is constant and so one obtains:

$$\ddot{r} = -\frac{GM}{r^2}$$

In General Relativity mass should not be regarded as constant when one goes from the energy conservation law to the equation of motion. The latter changes according to: $dM = -pdV$. The work of pressure forces changes mass or, equivalently, energy.

Let us insert $E = M$ into equation (9.8) :

$$\frac{dM}{dt} = -p\frac{dV}{dt} = -4\pi r^2 p\frac{dr}{dt}.$$

The equation of motion is obtained by differentiating eq.(9.15) having in mind the above relation:

$$\ddot{r} = -\frac{4\pi r^3}{3}(\rho + 3p)\frac{G}{r^2}.$$

The mass M' in this expression is different from the mass M:

$$M' = M + 4\pi r^3 \rho$$

or

$$M' = \frac{4\pi}{3}r^3(\rho + 3p)$$

In a medium with the equation of state $p = -\rho$ the mass M' is negative, $M' = -8\pi r^3 \rho/3$. Therefore gravitational attraction turns into gravitational repulsion:

$$\ddot{r} = \frac{8\pi G \rho}{3} r = H_0^2 r$$

The solution of this equation is:

$$r(t) = r_1 \cosh H_0 t + r_2 \sinh H_0 t$$

where r_1 and r_2 are arbitrary constants. The gravitational repulsion forces push particles away during inflation and later they move by inertia. This is how the Hubble expansion:

$$u = Hr$$

originates.

It should be noticed that there is a difference between a bomb explosion and the Big Bang. The explosion of a bomb is caused by a pressure gradient. There is no pressure gradient in the isotropic and homogenous Universe with equation of state $p = -\rho$. Negative pressure corresponds to stresses in a solid state. A pressure gradient appears on the boundary of a piece of matter with negative pressure as shown in figure 9.4. Pressure forces tend to squeeze this piece of matter. The repulsive forces in the Universe filled by matter with the equation of state $p = -\rho$, are caused by the fact that negative pressure changes the sign of the source of the gravitational field. Thus the push which caused the expansion of the Universe was the antigravitation created by negative pressure which existed in the very early Universe.

9.5 Quantum description of universe birth

Let us now turn to the quantum creation of the Universe. There was no classical spacetime in the initial state and the metric $g_{\mu\nu}$ was a purely quantum quantity. A good analogy which helps to understand the phenomenon of the transition from quantum spacetime to a classical one is α decay. Prior to the decay a notion of α particle trajectory was senseless, and its motion was a quantum one. After

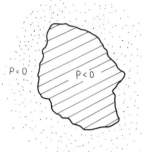

Figure 9.4: Compression of a piece of matter with negative pressure.

the channel transition α particle has left the nucleus and its trajectory with a good accuracy becomes the classical one. The birth of the Universe can be described in the same way. The classical spacetime was formed after a quantum tunnelling process and only after that can one talk about space and time , as we understand it, and about the creation of the Universe. Of course the theory of the creation of the Universe from nothing is not completed yet and so the following considerations are only illustrative.

The basic idea of a Universe created from nothing is that the probability of quantum production of a closed spacetime is nonzero. The probability of a classically forbidden process realized by a tunnelling transition in quantum mechanics is:

$$W = W_0 \exp(-S)$$

where S is the action and the preexponential factor W_0 can sometimes be obtained by exact calculations. The most important factor is $\exp(-S)$. The action S is calculated on classical trajectories in imaginary time i.e. on solutions of classical equations of motion $m\ddot{x} + u'(x) = 0$ where time is changed as $t \rightarrow i\tau$. That is the standard procedure to obtain the semi-classical approximation.

The probability of tunnelling in quantum field theory is calculated in the same way. Moreover, quantum field theory is often formulated in imaginary time for the purpose of ensuring the convergence of the functional integral:

$$Z = \int D\phi \exp(-S)$$

where $D\phi$ means the integration over all the states of quantum fields. If quantum states of gravitational fields are also considered the integration is done over all the metrics with signature $(+, +, +, +)$. The probability of a process is determined by the value of Z. In the semi-classical approximation the integral is calculated on the extremal values of S which correspond to solutions of classical equations of motion in imaginary time.

The euclidean action for gravitational fields interacting with a matter field ϕ is:

$$S = -\frac{m_{Pl}^2}{16\pi} \int \left[R - 2\Lambda + \frac{16\pi}{m_{Pl}^2} L(g_{\mu\nu}, \phi) \right] \sqrt{g} \, d^4x$$

where the integration is done over the whole four-dimensional manifold. The action is finite only if the manifold is closed, as it is for example of the de Sitter space with the metric:

$$ds^2 = -dt^2 + H^{-2} \cosh^2 Ht(dr^2 + sin^2 r d\Omega^2)$$

We calculate the action making the substitution $t \to it$. The metric becomes:

$$ds^2 = dt^2 + H^{-2} \cos^2 Ht(dr^2 + sin^2 r d\Omega^2) \tag{9.16}$$

It is the metric of a 4-dimensional sphere in 5-dimensional euclidean space. This can be easily checked looking at the metric of a spherical surface in 5-dimensional space (see sec.2 of chapter 5):

$$t = r_0 \sin H\tau; \quad v = r_0 \cos H\tau \cos r; \quad z = r_0 \cos H\tau \sin r \cos \theta$$

$$y = r_0 \cos H\tau \sin r \sin \theta \sin \phi; \quad x = r_0 \cos H\tau \sin r \sin \theta \cos \phi;$$

$$r_0 = H^{-1} = const;$$

$$g_{00} = 1; \quad g_{11} = H^{-2} \cos^2 Ht; \quad g_{22} = H^{-2} \cos^2 Ht \sin^2 r$$

$$g_{33} = H^{-2} \cos^2 Ht \sin^2 r \sin^2 \theta; \quad \sqrt{g} = H^{-3} \cos^3 Ht \sin^2 r \sin \theta$$

The 4 dimensional volume of this hypersurface is given by the expression:

$$V = H^{-4} \int_{-\pi/2}^{\pi/2} d(Ht) \int_0^\pi dr \int_0^\pi d\theta \int_0^{2\pi} d\phi \cos^3 Ht \sin^2 r \sin\theta =$$

$$= \frac{8}{3}\pi^2 H^{-4}$$

The curvature tensor invariant for metric (9.16) is given by:

$$R = -6a^{-3}\left[a - a\left(\frac{da}{dt}\right)^2 - a^2\frac{d^2a}{dt^2}\right]$$

where $a = H^{-1}\cos(Ht)$. This gives $R = -12H^2$. Expressing the cosmological constant Λ through H, one can obtain for the action in the case of empty space $(L(g_{\mu\nu}, \phi) = 0)$:

$$S = \pi H^2 m_{Pl}^2 \int \sqrt{g}\, d^4x = \pi m_{Pl}^2 H^{-2}$$

The probability of the formation of a Universe with radius H^{-1} is thus proportional to:

$$W \sim \exp\{-\pi m_{Pl}^2 H^{-2}\}$$

The probability is small when the radius is big $H^{-1} \gg m_{Pl}^{-2}$. The most probable is the creation of a universe with the radius which is of the order of the Planck one. The behavior of the scale factor $a(t)$ after creation of the Universe from nothing is illustrated in figure 9.5. The region at $t < 0$ is classically forbidden and is called nothing. The spacetime in this state is described by quantum equations. There is no classical spacetime there exactly as there is no classical trajectory of an α particle in the forbidden region in α decay. The scale factor and metrics undergo strong quantum fluctuations there.

Let us write the Schrödinger equation, which we may hope, describes the birth of the Universe. The energy of the Universe is:

$$\left(\frac{1}{a^2}\frac{da}{d\eta}\right)^2 - H_0^2 + \frac{k}{a^2}$$

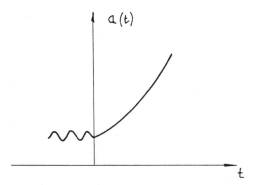

Figure 9.5: Evolution of the scale factor during the quantum birth of the Universe.

where $H_0^2 = 8\pi G\rho/3$. The corresponding Hamiltonian is:

$$H = \left(\frac{da}{d\eta}\right)^2 - H_0^2 a^4 + ka^2$$

Since the total energy of the closed Universe is equal to zero the Hamiltonian is also equal to zero. The quantization procedure consists of choosing a coordinate and momentum satisfying the commutation relation:

$$[q, p] = i$$

The scale factor can be chosen as the coordinate. The corresponding momentum is:

$$p = \frac{\partial H}{\partial(da/d\eta)} = \frac{da}{d\eta}$$

In terms of these variables the Hamiltonian can be rewritten as:

$$H = p^2 - H_0^2 a^4 + ka^2$$

With coordinate and momentum operators substituted into H the Schrödinger equation in coordinate representation can be written as:

$$H\psi = 0$$

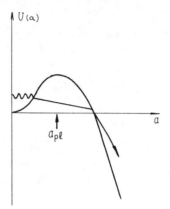

Figure 9.6: The effective potential that determines the dynamics of the scale factor.

It must be stressed that ψ depends on a only, and does not depend on time. There is no time dependence because the energy of a closed universe is equal to zero. It follows from the standard expansion of wave function in energy eigenstates:

$$\psi(a,t) = \sum_n \phi_n(a) e^{iE_n t}$$

The Hamiltonian consists of two terms: the kinetic one $\partial^2 \psi / \partial a^2$ and the potential one $V(a)\psi(a)$. In our case the potential is $V(a) = -H_0^2 a^4 + k a^2$. The ψ oscillations around zero do not correspond to the expansion. The Universe born with $a \leq a_{Pl}$ must cross the classically forbidden region (see figure 9.6). After the tunnelling transition (as in the case of α decay) it starts to expand exponentially.

Let us solve this equation for large a. We shall look for a solution in the form:

$$\psi(a) = \exp\{if(a)\}$$

with the condition:

$$f \frac{d^2 f}{da^2} \ll \left(\frac{df}{da}\right)^2$$

The equation for $f(a)$ is:

$$\left(\frac{df}{da}\right)^2 = H_0^2 a^4 - ka^2$$

The solution for large a i.e. for $a \gg a_{Pl}$ is:

$$f(a) = H_0 a^3 / 3$$

The wave function in the semi-classical region is:

$$\psi(a) = \exp\{iH_0 a^3 / 3\}$$

Let us find how the scale factor depends on time in this case. The time is introduced with the help of the momentum p:

$$p = -\psi^* i \frac{\partial}{\partial a} \psi = A_0 a^2$$

By definition $p = da/d\eta$ and $d\eta = dt/a(\eta)$ so $(1/a)(da/d\eta) = da/dt = H_0 a$, and we obtain:

$$a(t) = a_0 \cosh Ht$$

In this simple model the newly-born Universe starts to expand according to the de Sitter law.

9.6 Chaotic inflation and the eternal universe

Let ϕ be a randomly distributed scalar field in the Universe. The only condition we impose on it, is that its energy density is not higher than the Planck energy density:

$$\rho_\phi < m_{Pl}^4 \tag{9.17}$$

As it has been shown by Linde (1983) if there are regions where the ϕ field is homogeneous enough then they will undergo exponential expansion and occupy the main part of the volume of the physical Universe, but of course not the coordinate volume.

Let us consider for simplicity the case of a non-interacting field. The equation of motion of the field in the region of homogeneity, where the space derivatives are neglected, is:

$$\ddot{\phi} + 3H\dot{\phi} + m^2\phi = 0. \tag{9.18}$$

The Hubble parameter is expressed through the energy density as:

$$H^2 = \frac{8\pi}{3m_{Pl}^2}\left(\frac{\dot{\phi}^2}{2} + \frac{m^2\phi^2}{2}\right) \tag{9.19}$$

When the field is strong enough i.e. $\phi > m_{Pl}$ then equations (9.19) and (9.18) have the solutions:

$$\phi(t) = \phi_0 - \frac{m\, m_{Pl}\, t}{2\sqrt{3\pi}} \tag{9.20}$$

$$a(t) = a_0 \exp\left(\frac{2\pi}{m_{Pl}^2}\left[\phi_0^2 - \phi^2(t)\right]\right)$$

The scale factor rises exponentially until $t \approx 2\sqrt{3\pi}\phi_0/mm_{Pl}$. Once more we encounter the mechanism of inflation due to slow rolling down of field ϕ to the equilibrium. This mechanism of inflation is called chaotic inflation since the initial state is chaotic.

Long wave fluctuations of the field ϕ grow during inflation. They reach the value:

$$|\delta\phi| \approx H/2\pi \approx \sqrt{1/3\pi}\, m\phi_0/m_{Pl} \tag{9.21}$$

in a time $t \approx H^{-1} \approx \sqrt{3}\, m_{Pl}/2\sqrt{\pi}m\phi_0$. It is necessary that $m \leq 10^{-5}m_{Pl}$ to avoid contradiction with observations (see chapter 10 and 11). This means that ϕ_0 can reach a value of up to $10^4 m_{Pl}$ without violating the bound (9.17).

Quantum fluctuations of ϕ in the expanding Universe lead to a surprising phenomenon. Namely, the value of ϕ may grow despite the rise of the potential energy. The necessary condition for this can be found by comparing the change of ϕ according to (9.20):

$$\delta\phi = m_{Pl}^2/4\pi\phi$$

with the change of ϕ because of fluctuations (9.21). Hence it follows that if:

$$\phi \geq \frac{m_{Pl}}{3}\left(\frac{m_{Pl}}{m}\right)^{1/2} \approx 100 m_{Pl}$$

the field grows with a probability of about $\frac{1}{2}$.

An interesting scenario of the evolution of the Universe has thus been obtained. In a universe with randomly distributed field ϕ the regions of its sufficient homogeneity grow exponentially. The region of the size H^{-1} grows e times and its volume grows e^3 times in the period of H^{-1}. Thus, it turns into e^3 regions of the size H^{-1}, each continuing to expand exponentially. However, the field ϕ decreases in one half of them by $|\delta\phi| + \Delta\phi$ and increases by $|\delta\phi| - \Delta\phi$ in the other. Although the mean value of ϕ in coordinate volume falls according to (9.20) the physical volume of the regions with growing ϕ increases exponentially. In the regions where the field becomes smaller than $100 m_{Pl}$ the exponential expansion stops because ϕ starts to decrease there. Thus we arrive at the infinite process of creating universes, like ours, out of fluctuations of the field ϕ (Linde 1986). It is possible that there was no birth of the Universe as a whole. The process might have never started and will never end.

The size of the Friedman region is presently about :

$$m_{Pl}^{-1} \times \exp(2\pi\phi_0^2/m_{Pl}^2) \approx 10^{10^5}$$

i.e. its boundary is far beyond the horizon. Therefore there is no way to check this scenario but it is very interesting from the philosophical point of view.

Let us note that in this model the Universe as a whole is nonhomogeneous and its topology differs from the topology of the simple Friedman model. However at the scales of the present-day horizon the topological classification given above is of course valid and in particular if locally $\Omega > 1$ then our part of the Universe may collapse.

Chapter 10

Structure of the universe

The Universe is homogeneous only in the mean on large scales $L > 200$ Mps. On smaller scales, one sees prominent hierarchical structure starting from planets and stars to small globular clusters (1 ps), galaxies (10–100 kps), clusters of them (10 Mps) and superclusters (100 Mps). The latter form thin (10–20 Mps) walls surrounding vast empty spaces. Strictly speaking, these voids are the regions with size 100–200 Mps in which the density of luminous matter is much smaller than average. This makes the bubble structure of the Universe. No inhomogeneities have been found on bigger scales. Apart from the distribution of the luminous matter the observations of microwave and X ray background indicate that the Universe is homogeneous on scales $10^3 - 10^4$ Mps with 1 percent accuracy.

It is assumed that the hierarchical structure of the Universe is a result of gravitational instability of small initial density fluctuations. The origin of these fluctuations is discussed in the next chapter. They are connected with the physical processes in the early (inflationary) stage of the evolution of the Universe and thus their values are determined by fundamental constants of particle physics.

In this chapter the primordial density fluctuations will be taken as initial conditions. We discuss their amplitude, spectrum and evolution, necessary to obtain the present structure of the Universe. The physical properties of dark matter, that probably determine the structure formation are also discussed.

First we shall consider the evolution of inhomogeneities in a Universe filled with baryonic matter and radiation only. The Universe containing many forms of matter will be discussed later on.

This chapter is based on the article by Ya.B.Zeldovich [73] and on the book by P.Peebles [28].

Since the structure of the Universe is not the main topic of this book our presentation here is rather brief and refers mainly to the physics of the very early Universe.

10.1   The initial period of growth of fluctuations

E.M.Lifschitz in 1946 [22] formulated a relativistic theory of the evolution of perturbations. Qualitatively his results can be understood in the framework of Newtonian gravity. Gravitational attraction is the physical reason for the growth of fluctuations. Let us consider a region with density excess $\delta\rho$ in the homogeneous infinite space filled with matter with density ρ_0. This region will attract surrounding matter. The density inhomogeneity will grow if the pressure forces are smaller than gravitational ones. Gravitational forces dominate if the region with density excess is large enough, and pressure forces dominate when it is small. The scale that divides these two possibilities can be found as follows. The time of the free fall to the centre of the inhomogeneity is $t_g \approx (G\rho_0)^{-1/2}$ while the characteristic time of sound wave propagation is $t_s \approx L/c_s$. The fluctuations grow if $t_g < t_s$. The fluctuations oscillate and form sound waves if the opposite case $t_s > t_g$ is true. Hence only the fluctuations on scales $L > c_s(G\rho_0)^{-1/2}$ can grow. The limiting length is called the Jeans length λ_J. Jeans considered the problem of gravitational instability in Newtonian gravity at the beginning of the century. The exact value of λ_J differs from that given above by a numerical factor:

$$\lambda_J = \sqrt{\frac{\pi}{G\rho}}\, c_s \qquad\qquad (10.1)$$

It is shown in refs.[28,73] that the fluctuations of wavelength λ and amplitude A_k evolve according to:

$$\frac{\delta\rho}{\rho_0} = A_k e^{\gamma t + ikx} \tag{10.2}$$

where $k = 2\pi/\lambda$ and $\gamma = \pm\sqrt{4\pi G\rho_0 - k^2 c_s^2}$.

The growing solution can be factorized for small $\delta\rho$ and k:

$$\delta\rho = \rho_0 e^{\gamma_0 t} \int A_k e^{ikx} d^3x = \rho_0 e^{\gamma_0 t} \psi(x) \tag{10.3}$$

where $\gamma_0 = \sqrt{4\pi G\rho_0}$. Taking into account the higher order terms results in the substitution of $\psi(x,t)$ instead of $\psi(x)$.

Density fluctuations in stationary background grow exponentially. This growth is slower in the expanding Universe. The density is a decreasing function of time:

$$\rho_0(t) = (6\pi G t^2)^{-1} \tag{10.4}$$

(this formula is valid for $\Omega = \rho/\rho_c = 1$). The fluctuation growth can be naively estimated to be:

$$\exp\left(\int^t \sqrt{4\pi G\rho_0(t)}\, dt\right) = \exp\left(\sqrt{\frac{2}{3}} \int \frac{dt}{t}\right) \approx t^{\sqrt{2/3}}$$

A power law was obtained instead of an exponential one. The exact result, based on hydrodynamics in an expanding Universe, is slightly different:

$$\frac{\delta\rho}{\rho_0(t)} = t^{2/3}\psi(x). \tag{10.5}$$

The relative value of fluctuations grows with time as the scale factor (in the matter dominated Universe with $\Omega = 1$). This result was obtained for small fluctuations when a linear approximation is good. The strict relativistic theory of evolution of small fluctuations was formulated by Lifschitz. It can be found in the book by L.Landau and E.Lifschitz [20]. For the sake of brevity we present here only their results. The metric tensor fluctuations can be expressed as a sum of three terms. The first one is the scalar fluctuations described above. In this case the correction to the metric tensor can be expressed by one scalar function. Scalar fluctuations are the most important for the formation of the structure of the Universe since already in the first order they lead to matter density and velocity fluctuations. Scalar fields are connected with the divergence $\partial_\mu v^\mu$ of the velocity field v_μ of the matter. The second type of fluctuation is connected with the antisymmetric part of the tensor $\partial_\mu v^\nu$. This is vector or vortex fluctuation. It corresponds to rotation of matter i.e to velocity perturbations. In this case density fluctuations arise only in the second order of perturbation. Tensor fluctuations of metric are connected with the symmetric part of $\partial_\mu v^\nu$. As in the former case the matter density is not affected by these fluctuations in the first order. They correspond to gravitational waves. The importance of a search for relic gravitational waves was understood recently, because they should have increased very much during the inflationary stage, and their discovery could give a lot of information about the inflationary period. This will be discussed in chapter 11. Scalar fluctuations of the metric are constant for small t, and for wave-length smaller than the horizon during the radiation dominated era. The corresponding relative density fluctuations increase with time as $\delta\rho/\rho_0 \approx t$ since $\rho_0 \sim t^{-2}$ and $\delta\rho \sim t^{-1}$. Thus it is convenient to describe the initial fluctuations in terms of metric fluctuations. In the matter dominated Universe the fluctuations start to grow as the scale factor in accordance with expression (10.5). The amplitude of the vector mode is singular for small t i.e $\delta h_{\mu\nu} \to \infty$ when $t \to 0$. This follows from the angular momentum conservation law. Hence no weak vector modes exist.

10.2   Fluctuations in plasma

The preceding section was concerned only with one component matter.The primordial plasma contained protons, electrons, electromagnetic radiation, neutrinos or maybe some other particles that form the hidden mass. Thus density fluctuations could not be described by the simple quantity $\delta\rho$ but the fluctuations of each component of plasma $\delta\rho_i$ should be taken into account. Postponing discussion of invisible matter to sections 5 and 7 we consider here density fluctuations in plasma containing electromagnetic radiation and massive ionized matter. These fluctuations can be divided into two independent modes, adiabatic and isothermal. In the first case the fluctuations of specific entropy are zero and fluctuations of energy density of matter and radiation are related by:

$$\frac{\delta \rho_m}{\rho_m} = \frac{3}{4} \frac{\delta \rho_r}{\rho_r}$$

It means that the ratio of number densities does not fluctuate:

$$N_m / N_r = const$$

Adiabatic fluctuations arise from the initially homogeneous distribution of matter, due to scalar fluctuations of metric. Fluctuations of the metric lead to different expansion rates in different spacetime points and thus lead to density fluctuations. Chemical contents i.e. the ratio N_m / N_r is not altered since the matter–radiation interaction is strong.

Isothermal fluctuations are defined by the condition $\delta \rho_r = 0$. Radiation is homogeneous, there are no temperature fluctuations and matter density changes slightly from one point to another, $\delta \rho_m \neq 0$. Isothermal fluctuations are determined not only by initial values of metric fluctuations and also by the distribution of chemical content in the primordial plasma.

The theory presented in this chapter is supposed to describe the structure of the Universe given the spectrum of initial inhomogeneities and the physical properties of density fluctuations. The structure formation is different depending on the kind of fluctuations chosen, because adiabatic and isothermal fluctuations evolve differently.

It is convenient to divide the fluctuations into isothermal and adiabatic modes, since they do not interact and evolve separately in a rather wide range of wavelengths. Adiabatic fluctuations remain adiabatic for long wavelengths when entropy is conserved, and photon diffusion can be neglected. There is no diffusion of photons when the fluctuations are isothermal. Isothermal fluctuations may become adiabatic only due to gravitational interactions. This effect is small for wavelengths smaller than the horizon, since the Jeans length is about the horizon size for relativistic matter.

In what follows we consider the development of the structure of the Universe in the cases of adiabatic and isothermal fluctuations. These two cases are significantly different.

Adiabatic fluctuations seem natural from the physical point of view but one must keep one's eyes open to the possibility of isothermal fluctuations. It is possible that both modes are necessary for the description of the structure of the Universe.

10.3   Notes on the fluctuation spectrum

The fluctuations can be described by their spectral distribution. Let us consider scalar fluctuations of metric in the synchronous background $ds^2 = dt^2 - a^2(t)(dx^2 + dy^2 + dz^2)$ and let us assume that they are random with vanishing mean value $\langle h(x) \rangle = 0$. The spatial distribution of fluctuations can be described by their correlation function:

$$G(x, y) = \langle h(x)h(y) \rangle$$

which is the mean value of the product of fluctuations at space points x and y. It is usually assumed that because of the Universe's homogeneity fluctuations are translationally invariant:

$$G(x, y) = G(x - y).$$

The spectrum is determined by the Fourier transform of $G(x)$:

$$h_k^2 = \int d^3x\, e^{ikx} G(x). \tag{10.6}$$

The simplest and physically reasonable assumption about h_k is that the fluctuations are scale-independent. Dimensional analysis gives:

$$h_k = \delta k^{-3/2}, \tag{10.7}$$

where δ is a dimensionless constant (the dimension of $h(x)$ is different than that of h_k).

When it is assumed that $h_k \approx k^{-\nu}$ then astronomical observations indicate that $\nu \approx 3/2$. If $\nu > 3/2$ then the Universe would be less isotropic on large scales than observed. If $\nu < 3/2$ then the density inhomogeneities on small scales would be too big and too many black holes would form. They could be observed through their X-ray radiation. The constant δ should be about 10^{-4} lest there is a contradiction with isotropy of microwave background radiation.

The spectrum $h_k \sim k^{-3/2}$ is the so-called fractal spectrum. The importance of fractals in physics was understood only very recently thanks to the book by Mandelbrot (1977).

The physical meaning of fractals can be described with the example of measuring the length of a shore line on different scales. The shore line is a smooth line on low resolution maps but more and more details appear with better resolution. One could distinguish small bays, mouths of streams and even separate stones. With even better resolution there appear small scrapes on the surface of the stones and even the crystal structure. With further increase of resolution one enters the quantum world where the notion of boundary disappears. The length of the shore line changes with the resolution scale. It approaches infinity for infinitely small scales. Thus this length can be defined only if a scale is given.

For physicists fractal curves are closely connected with the notion of intermediate asymptotics. A curve with $h_k \approx k^{-3/2}$ is a fractal curve (Ya.B.Zeldovich, D.D.Sokolov 1985).

General properties of fractal curves will be illustrated by an example of a continuous function that has no derivatives:

$$y(x) = \sum_{n=0}^{\infty} a(k_n) \cos(k_n x + \phi_n)$$

where $a(k_n) \approx k_n^{-\alpha}$ and $k_n \approx n$ when $n \to \infty$. The phases ϕ_n are distributed randomly in the interval $(0, 2\pi)$. It is essential that the phases should be random and that k_n is proportional to n but not equal to it. In the latter case the fractal curve may turn into a smooth, differentiable curve except for a countable number of points where $y(x) = \infty$. In other words it is essential that not only phase but also frequencies are random. Such a curve is continuous but not differentiable when $0 < \alpha < 1$. Mathematical analysis of these curves is done with the use of the so-called Hölder derivatives:

$$\Delta y = \mu(\Delta x)^\alpha$$

An example of such a derivative occurs in the study of Brownian motion for which:

$$\Delta l = \sqrt{D\Delta t},$$

that is the Hölder index is $\alpha = 1/2$.

Now let us define the dimension of a fractal curve. The Hausdorf algorithm for definition of the dimension goes as follows. Let us draw circles with radius $\epsilon \to 0$ around each point in the curve and calculate the surface covered by all these circles $S(\epsilon)$. The rate of change of $S(\epsilon)$ with respect to ϵ determines the dimension. For a smooth curve it is $S(\epsilon) \sim \epsilon L$ where L is the curve length. It is $S(\epsilon) \sim \epsilon^2$ for a point and $S(\epsilon) \approx \epsilon^0$ for a plane.

The dimension of a fractal curve (10.7) will be calculated in the same way. The characteristic amplitude of oscillations on the scale ϵ is $a(\epsilon^{-1})$. If $a(\epsilon^{-1})$ increases slower than ϵ as $\epsilon \to 0$ then the surface covered by all these circles is not of the order of ϵ but is about $a(\epsilon^{-1})$. If $a(\epsilon^{-1}) \leq \epsilon$ then the surface follows all the zigzags of the curve and the curve is smooth.

For a fractal curve $S(\epsilon) \sim a(1/\epsilon) \approx \epsilon^\alpha$. If $0 < \alpha < 1$ $S(\epsilon)$ decreases slower than for a smooth line but faster than for a plane. Thus fractal curves occupy intermediate position between smooth curves and 2-dimensional surfaces. That is why fractal curves are called sometimes thick lines. Dimension of a fractal curve was defined by Hausdorf as:

$$dim_{ext}\gamma = 2 - \alpha$$

where the index ext means that in the definition of the dimension the external space around the curve is essential.

Upper and lower cutoffs are usually imposed on the spectrum of fractal curves in physics. The longwave cut-off corresponds to the time of observation and the shortwave one to the accuracy of measurement.

The spectrum (10.7) gives a rather good description of the observed astronomical picture on large, as well as small scales. Such a spectrum of adiabatic fluctuations can be approximately obtained in a simple inflationary model (see chapter 11). A more detailed description of the Universe's structure requires introducing dimensional parameters already in the initial fluctuations. Note that an initially scale-independent spectrum with $\nu = 3/2$ is altered in the course of the expansion of the Universe. There appear different scales connected with damping of small waves with certain lengths, with mean free path of particles and so on.

Some additional scale parameters may also be needed. There are a lot of different scale parameters presented by microphysics. Unfortunately there are too many of them and the problem is not to find a physical mechanism leading to a scale-dependent spectrum but to make the right choice among many possibilities. The properties of the hidden mass are unknown and neither are the interactions of particles at high energy, so the theory of the Universe structure formation may be a source of information about them.

10.4 Structure of the universe in the adiabatic theory

At temperatures higher than the hydrogen recombination temperature $T = 3000K$ matter and radiation interact strongly and are thus tightly bound to each other. Metric fluctuations weakly depend on time when the wavelength is larger than the horizon $\lambda > t$. The density fluctuations grow as the scale factor in the nonrelativistic case, and as the scale factor squared in the relativistic case. This rise can be understood as a slower expansion of regions with density excess and correspondingly a faster expansion of regions with density deficit relative to average homogeneous expansion [1]. This is

---

[1]The fluctuations with size bigger than the horizon at this stage are especially interesting from the cosmological point of view because reentering the horizon

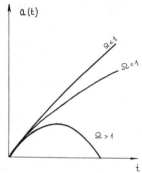

Figure 10.1: The time dependence of the scale factor for different values of $\Omega = \rho/\rho_c$. The bigger is the density the slower is the expansion.

illustrated in figure 10.1. The fluctuations start to oscillate forming sound waves when their size becomes smaller than the horizon (the Jeans wavelength in radiation-dominated plasma is about the horizon size). The shortwave fluctuations are damped by the photon diffusion (Silk damping). The minimum wavelength of the fluctuation that is not damped can be estimated as follows. The mean free path of photons in plasma is $l = (\sigma n_e)^{-1}$ where n_e is the electron number density and $\sigma = 2\pi\alpha^2/m_e^2$ is the cross section of photon-electron scattering (Thompson cross section), m_e is the electron mass, and $\alpha = 1/137$ is the fine structure constant. A photon diffuses a distance L in time $t(L) = L^2/l$. This time should not be larger than the age of the Universe $t(L) < t$. Hence the minimal wavelength of a fluctuation surviving to the moment of recombination t_r is:

$$\lambda_{min} \approx (l\, t_r)^{1/2}$$

and the minimum mass of the objects formed by this fluctuation is of the order of :

---

during the matter-dominated era, they lead to formation of the large scale structure of the Universe. Generation of perturbations with wavelength larger than the horizon looks troublesome. But here once more inflation helps. The wavelength of a fluctuation exponentially increases during the inflationary stage and goes far out of the horizon. So there is no contradiction with causality.

$$M_{min} = \frac{4\pi}{3}\rho_m \lambda_{min}^3 \tag{10.8}$$

where ρ_m is the matter density during recombination. This gives $M_{min} \approx 10^{14} M_\odot$ where $M_\odot = 2 \times 10^{33}$g is the mass of the Sun.

After recombination the conditions drastically change because the matter becomes electrically neutral. The radiation no longer interacts with matter. The velocity of sound in the matter falls down to about $(T/m_H)^{1/2}$ where T is the temperature and m_H is the hydrogen atom mass. The Jeans mass (i.e. the mass in a volume bounded by λ_J (10.1) is of the order of :

$$M_J^B = \frac{4}{3}\pi \lambda_J^3 \rho \approx 10^6 M_\odot$$

Fluctuations with wavelength larger than λ_J should increase after recombination. However because of dissipation of short waves no perturbation exists up to M_{min} (10.8). The evolution of fluctuations is illustrated in figures 10.2, 10.3 and 10.4.

For adiabatic perturbations the fluctuations of temperature are expressed through energy density fluctuations as:

$$\frac{\delta T}{T} = \frac{1}{3}\frac{\delta \rho}{\rho} \tag{10.9}$$

The energy density fluctuations after recombination grow as the scale factor (if $\Omega \approx 1$) but the temperature fluctuations remain unchanged. Thus measuring $\delta T/T$ today one may find the value of density fluctuations during recombination. The existing upper bounds on temperature fluctuations permit too small values of $\delta T/T$. The corresponding energy density fluctuations cannot reach the necessary value $\delta \rho/\rho \approx 1$ during the period from recombination to the present day (see section 9 in this chapter).

This problem can be solved by introducing dark matter. Another way of solving it is to take account of isothermal fluctuations. Both these possibilities will be discussed later on.

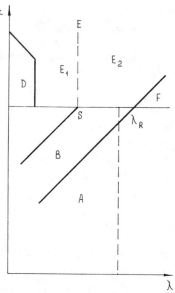

Figure 10.2: Different regions on the time-wavelength plane where the fluctuations evolve differently.

A- radiation dominated plasma, wavelength is bigger than horizon. Fluctuations of metric are constant while density fluctuations grow.

B- radiation dominated plasma, fluctuations smaller than horizon. Density fluctuations oscillate and form sound waves.

C- region of damping of acoustic fluctuations. The horizontal line at $t_R \approx 3 \times 10^5$ years corresponds to hydrogen recombination when matter and radiation decouple. The fluctuations with wavelength longer than S can survive till t_R. λ_H is the Jeans length of neutral hydrogen. λ_R is the Jeans length of ionized plasma.

D- region where fluctuations in neutral hydrogen do not increase.

E- region where fluctuations in neutral hydrogen grow. There are no fluctuations in E_1 due to damping in C but they would grow if they were there. The most important rising fluctuations are in E_2.

F- region of growth of long wave fluctuations in neutral hydrogen. The wavelength is bigger than horizon. The relative density fluctuations on the line between A and B or E_2 and F, where the wavelength is equal to horizon, are equal to initial metric fluctuations.

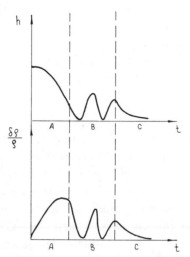

Figure 10.3: Evolution of shortwave fluctuations of metric h and density $\delta\rho/\rho$ as functions of time (shorter than S in figure 10.2). The letters A, B, C have the same meaning as in figure 10.2.

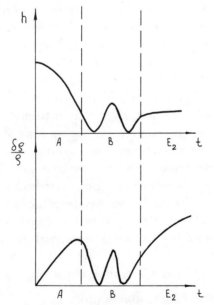

Figure 10.4: Evolution of long wave fluctuations (longer than S in figure 10.2). The letters have the same meaning as in the figure 10.2).

10.5   Neutrino dominated universe

Neutrinos may form the hidden mass of the Universe. Relic neutrinos are abundant. The contemporary concentration of each type of neutrino and antineutrino is:

$$N_\nu = N_{\bar\nu} = 75 \text{cm}^{-3}(T_0/3\text{K})^3$$

From the point of view of the theory it is natural that the neutrino mass is nonzero. The experimental situation is uncertain nowadays. One can definitely say that the mass of the electron neutrino is smaller than 30 eV. The positive results about nonvanishing neutrino mass found by the ITEP group are not yet confirmed or rejected by other laboratories. Cosmology gives an upper limit on stable neutrino mass of the same order of magnitude. Cosmology needs massive neutrinos with the mass near the upper limit. It is hardly probable that masses of ν_μ and ν_τ will soon be measured in laboratory experiments because the characteristic energy of ν_μ is about 50 MeV, and that of ν_τ an order of magnitude larger. There is a possibility that a detailed investigation of the solar neutrino spectrum with the resonance oscillations taken into account (Mikheev, Smirnov 1986 [24]), will allow one to measure the masses of all neutrinos but this is not a task for the near future.

Already at the moment when the primordial plasma temperature was $T = 3 - 5$ MeV neutrinos ceased to interact with matter. One could conclude that shortwave fluctuations of neutrino energy density should grow since there is no pressure which could prevent gravitational contraction. This is not true since neutrinos freely stream from large concentration regions just because of weakness of their interactions. The minimum wavelength of the fluctuation can be estimated as follows. It is equal to the length of the path that a neutrino travels until it becomes nonrelativistic due to the redshift:

$$\lambda_J^\nu \approx t(T = m_\nu) \approx m_{Pl}/m_\nu^2 \qquad (10.10)$$

The minimum mass of such fluctuations is:

$$M_J^\nu \approx \rho_\nu \lambda_J^{\nu^3} \approx \lambda_J^{\nu^3} m_\nu^4 = M_{Pl}^3/m_\nu^2 \qquad (10.11)$$

Numerically it is equal to $M_J^{\nu} \approx 10^{16} M_{\odot} (10\text{eV}/m_{\nu})^2$. This mass is also called the Jeans mass although its physical nature is quite different than that of the classical Jeans mass. The fluctuations of this size start to grow earlier than any other. They are the first to reach the relative value of the order of unity. So they form seeds to clusters of galaxies in this theory. The evolution of fluctuations in the neutrino model is illustrated in figure 10.5 (in logarithmic scale).

The temperature fluctuations can be an order of magnitude smaller in a neutrino dominated Universe ($m_{\nu} \approx 30\text{eV}$) than that in the baryon dominated Universe. The fluctuations of neutrino energy density start to grow when neutrinos become nonrelativistic i.e. at $z \approx 10^4$ (if $m_{\nu} \approx 30\text{eV}$). The baryon fluctuations start to grow at the moment of recombination when $z \approx 10^3$.

Fluctuations $\delta T/T$ originated at $z \approx 10^3$ and exist until today. The growth of neutrino energy density fluctuations from $z = 10^4$ to $z = 10^3$ has almost no impact on the radiation temperature fluctuations. Thus neutrino energy density fluctuations can be 10 times larger than that of baryons in the baryon dominated Universe with the same value of $\delta T/T$.

The Universe's structure manifests itself by ordinary baryonic matter only. Only this kind of matter forms stars and other celestial bodies that emit electromagnetic radiation which can be detected. Neutrinos, which might give 90 percent of the mass of the Universe are unobservable, but their clustering can be traced by the inhomogeneities in the ordinary matter which falls down into the gravitational potential well dug up by neutrinos. In this sense the observed structure of the Universe is secondary with respect to neutrino motion and clustering.

If the presented scenario is correct the role of neutrinos in the Universe is difficult to overestimate. This particle which escaped experimental detection for a such long time that it even shook belief in the energy conservation law probably makes 90 percent of the mass of the Universe and no galaxies, stars and planets with human beings can be formed without it.

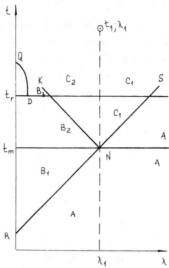

Figure 10.5: Evolution of fluctuations in neutrino dominated Universe. On RNS $\lambda = t$ (wavelength equal to horizon), NK is the Jeans wavelength for heavy neutrinos, QD is the Jeans wavelength of ordinary matter, A is the region where metric fluctuations are constant while density fluctuations (of any type of matter) grow till they reach the value

$$\left(\frac{\delta\rho}{\rho}\right)_\nu \approx \left(\frac{\delta\rho}{\rho}\right)_m \approx \left(\frac{\delta\rho}{\rho}\right)_R \approx h$$

on the RNS line; t_m is the time when neutrinos become nonrelativistic; NK is the line where $\lambda = v_\nu t_r$ (v_ν is the neutrino velocity); t_r is the time of hydrogen recombination. The neutrino fluctuations are damped in the regions B; in B_1 neutrinos are relativistic, B_2 corresponds to nonrelativistic neutrinos in radiation dominated plasma, B_3 corresponds to nonrelativistic neutrons in neutral hydrogen gas and radiation. In C_1 and C_2 neutrino fluctuations grow. The plasma does not move in B_2 and C_1. Neutral ordinary matter is attracted by gravitational field of neutrino in B_3 and C_2. t_1 is the time needed for the most slowly rising perturbation with the wavelength given by eq. (10.10) to reach the value of the order of unity. After t_r there appear fluctuations in ordinary matter due to gravitational attraction of neutrino fluctuations in C_2. Only the fluctuations with the wavelength $\lambda > \lambda_J^\nu$ survive till this moment because of the dumping in B_2 and B_3.

10.6 Pancakes and the bubble structure

The model presented above predicts the bubble structure of the Universe, where large voids are divided by relatively thin walls with higher density. This result is based on the assumption that the pressure is small and short wave fluctuations are dumped. It agrees with astronomical observations showing that clusters of galaxies form surfaces with a thickness of about 10–20 Mps surrounding empty regions with characteristic size of 100–200 Mps.

In cosmology the initial fluctuations are random. This leads to stochastic behavior of increasing density fluctuations. The probability of dilution or compression of matter along any axis is about $\frac{1}{2}$. The proper choice of the coordinate system may make this probability independent on the motion of matter along other axes. Hence, the probability of contraction in three axes simultaneously is $\frac{1}{8}$. This is of course a crude estimate. Nonlinear effects make it twice as small. The concentration of matter in the region of compression must be high. The velocity of contraction along different axes is different. Thus the motion down the fastest contraction axis leads to formation of flat pancake-like structures. These pancakes are later transformed to superclusters of galaxies.

The question is how does the bubble structure arise ? The probability of contraction along an arbitrary axis is $\frac{7}{8}$. In the gas with vanishing pressure the compression even along one axis leads to collision of paths of particles and to formation of an infinite density surface. Therefore about 90 percent of matter comes into the region with high density.

Let us illustrate this with a two dimensional example (figure 10.6). There are three types of regions: a, which contracts along two axes, b, which contracts along one and expands along the other axis, and c that expands in every direction. Regions a and c each fill 25 percent of the surface. They form islands in the sea of b.

Let us consider what happens because of particle motion. Regions a contract, regions b contract in one direction and expand in the other. So sea b transforms into a system of channels while

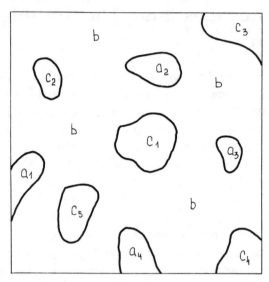

Figure 10.6: Initial configuration of regions with contraction in both axes: a_1, a_2, a_3, a_4; contraction in one axis and expansion in another: b; expansion in both axes:c_1, c_2 and c_3.

regions c grow in every direction enlarging the surface.

It is essential that the movement is continuous. Close points remain close, and closed curves remain closed. In particular closed curves dividing c and b remain closed. The structure thus obtained is shown in figure 10.7. It resembles the observed structure of the Universe.

The formation of pancakes can be described mathematically as follows. The spatial coordinate of a particle in a growing fluctuation mode is:

$$r(\xi, t) = a(t)\xi + b(t)\mathrm{grad}_\xi \psi(\xi) \qquad (10.12)$$

where ξ is the Lagrange coordinate, $a(t)$ is the scale factor, $a(t) \sim t^{2/3}$ in the nonrelativistic expansion regime for $\Omega = 1$. In the increasing mode $b(t)$ grows faster than the scale factor $b(t) \sim t^{4/3}$. The first term in equation (10.12) describes nonperturbed movement of particles, and the second one describes an increasing fluctuation. The function $\psi(\xi)$ is smooth because high frequency components

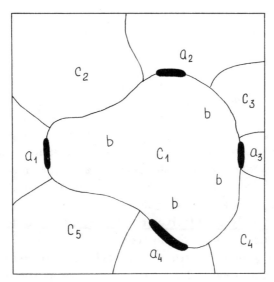

Figure 10.7: The result of evolution of the initial state shown in figure 10.6. Caustic surfaces are formed.

are damped in relativistic plasma.

Matter density can be found with the relation:

$$dm = \rho d^3 x = \rho_0 d^3 \xi \qquad (10.13)$$

or $\rho = \rho_0 / \det[\partial r / \partial \xi]$ where $\det[\partial r / \partial \xi]$ is the Jacobian of the transformation from ξ to r.

The direction of motion of each particle can be chosen so that the mixed derivatives disappear:

$$\frac{\partial^2 \psi}{\partial \xi_1 \partial \xi_2} = \frac{\partial^2 \psi}{\partial \xi_2 \partial \xi_3} = \frac{\partial^2 \psi}{\partial \xi_3 \partial \xi_1} = 0$$

Denoting the remaining three second derivatives as:

$$\frac{\partial^2 \psi}{\partial \xi_1^2} = -\alpha$$

$$\frac{\partial^2 \psi}{\partial \xi_2^2} = -\beta$$

$$\frac{\partial^2 \psi}{\partial \xi_3^2} = -\gamma$$

one obtains:

$$\rho(t) = \frac{\bar{\rho}(t)}{(1 - \frac{b}{a}\alpha)(1 - \frac{b}{a}\beta)(1 - \frac{b}{a}\gamma)}. \tag{10.14}$$

Let the axes be chosen so that $\alpha > \beta > \gamma$. Thus, the density is infinite when $\alpha b(t)/a(t) = 1$. Since α, β and γ are generally different, only one factor in the denominator vanishes. This means that compression goes only along one axis and thin high density surfaces of matter are thus formed.

These are only qualitative considerations. The detailed mathematical theory is beyond the scope of this book. Note that the formation of bubble structure in a stochastic process can be easily seen on a sunny day under a bridge, looking at the light patterns reflected by water.

The scenario we discussed is far from being complete. Neither formation of stars nor galaxies from gaseous matter has yet been discussed. The theory indicates general properties of the distribution of matter. This distribution is determined by motion of particles when pressure is negligible, and thus shortwave fluctuations are suppressed. This picture agrees with adiabatic fluctuations during the RD period. Evidently this scenario can be realized in the neutrino dominated universe.

The problem of background temperature fluctuations is milder in cosmology with massive neutrinos. This fact together with formation of the bubble structure are weighty arguments in favor of this scenario. But one should not place all the stakes on one horse, so later we discuss other possible types of hidden mass and another form of fluctuation.

The bubble structure of the Universe is gravitationally unstable and so this is not the final stage of the evolution. Pancakes decay into separate galaxies or form "strings", and knots. The later evolution should lead to large unbound clusters. The picture we observe in the sky now is only an intermediate stage of the evolution. It is the so-called intermediate asymptotics.From this point of view our

Universe is still very young.

10.7 Structure of the universe in the models with hidden mass

The neutrino seems to be the most natural candidate to form the hidden mass. If, moreover, the ITEP result that neutrino mass is about 20 eV is confirmed, then it will be established that the dark matter is in fact formed by neutrinos. However at the moment we are not sure that other possibilities are excluded. In contrast to the neutrino which does exist, other possible bearers of hidden mass are only hypothetical. Their existence is, with larger or smaller probability, predicted (or maybe it is better to say "permitted") by the theory but they have never been seen in experiment. Their general properties are that either they interact very weakly, or that their masses are too large to observe them in laboratories.

Independently of the detailed properties of the particles the dark matter can be divided into three types: cold, warm and hot. Particles which were in thermal equilibrium with the primordial plasma down to $T < 100$ MeV form hot dark matter. Their number density is close to the number density of relic photons. Their mass is bounded to a few tens of electronovolts. The name "hot" corresponds to the fact that they were relativistic at the moment of the decoupling. They cool down with the expansion of the Universe and at a certain moment become nonrelativistic. If their energy density is higher than that of baryons, they start to dominate the Universe. The hot hidden mass has already been discussed in the previous section. The main result is that fluctuations with size smaller than that of clusters of galaxies are suppressed (see eqs.(10.11) and (10.12)).

In the case of cold dark matter, the minimum size of fluctuations is smaller than galaxy size. The perturbation damping due to free streaming of these particles is not cosmologically significant. A heavy neutral lepton with mass greater than a few GeVs, according to the well known cosmological bounds, may form such mass. These particles are nonrelativistic at the decoupling time and so they are called cold. The concentration of relic heavy leptons should be at least 10^8 times smaller than the concentration of relic neutrinos.

The most popular possibility for the cold hidden mass is the coherent oscillations of the axion field. The axion is a hypothetical particle that ensures CP conservation in strong interactions. The original model with a relatively heavy axion ($m_a = 0.1 - 1$ MeV) was ruled out experimentally. Now the models with very light ($m_a = 10^{-5}$ eV) almost sterile axion are considered. The particle with these properties can have an impact on cosmology.

The warm dark matter is interspersed between hot and cold. In this case the damping of fluctuations starts just at the characteristic galactic size. Particles forming warm dark matter could have mass of about 1 keV and interact much more weakly than a neutrino. Their concentration should be a few orders of magnitude smaller than that of relic photons.

Relativistic matter still dominates the Universe when the bearers of warm (as well as cold) hidden mass become nonrelativistic. This is the essential difference with the hot dark matter case. Fluctuations of relativistic matter density with wavelength smaller than the horizon size, do not increase but they oscillate during this period. Since the energy density of dark matter at this stage is smaller than that of relativistic matter the fluctuations in hidden matter freeze and start to increase again in the matter dominated period.

In the model with warm dark matter, galaxies and small clusters of them, are formed first, and structures of large size are formed later.

A satisfactory description of large scale structure of the Universe can be obtained in the cold or warm dark matter model but for the detailed results a more complicated theory is needed.

The inflationary model of the Universe predicts $\Omega = 1$ but astronomical observations indicate that $\Omega = 0.2 - 0.3$. One can conclude that there exists a homogeneous background with $\Omega = 0.7 - 0.8$ which cannot be observed directly. This background could be formed by decays of long-lived hidden mass particles ($\tau = 10^5 - 10^6$ years) or by the cosmological constant. This multicomponent model presents a better description of the structure of the Universe than e.g. the simple neutrino model.

Discussing the qualitative description of the Universe's structure we have not yet explained how it is done. This problem is by no means a trivial one. One way is to use the method of the correlation function $\overline{\rho(x)\rho(y)}$. The characteristic size of cosmic clusters can thus be established but nothing can be said about their form or shape. A beautiful method proposed by S.F.Shandarin permits one to avoid this difficulty. It is based on the notion of percolation and goes as follows. Let us consider a set of material objects somehow distributed in space and draw a sphere with radius R around each object. All the objects that are within one sphere will be called unified. Let us increase R until the unified objects form chains passing through all the space. The critical value corresponding to this moment is denoted R_c. The characteristic dimensionless parameter is the number of objects inside a sphere with radius R_c:

$$N = \frac{4\pi}{3}R_c^3 n \qquad (10.15)$$

where n is the average density of the objects.

Let R_{c0} and N_0 correspond to a homogeneous distribution of objects. The values corresponding to separate clusters are larger, and that corresponding to the bubble structure are smaller. Thus R_c and N deviate from the homogeneous case in different directions for clustered objects and for the bubble structure. The analysis of the distribution of galaxies gives evidence in favor of the bubble structure of the Universe. This is typical for adiabatic fluctuations and contradicts the simplest form of isothermal fluctuations.

10.8 Isothermal fluctuations

The temperature and density of radiation is homogeneous in the case of isothermal fluctuations. Only the density of matter is disturbed. For such perturbations all the wavelengths up to the horizon are not damped during the radiation-dominated era. Moreover, there is a specific mechanism of fluctuation growth in this model. The regions with baryon excess cool down faster than those with baryon deficit because the radiation is transmitted to nonrelativistic particles. Radiation diffuses to the lower temperature regions carrying baryons along. Thus the fluctuations in ρ_B become larger.

After recombination the fluctuations with wavelength larger than λ_J (10.1) still increase due to gravitational instability, while the fluctuations with smaller wavelength oscillate and are quickly damped by photon diffusion. The fluctuations with the scale of about λ_J are the first to reach the relative value of the order of unity and form objects with characteristic mass $\approx 10^6 M_\odot$. This is the typical mass of a globular cluster. Larger structures are formed later because of clustering of such nebulae contrary to the adiabatic theory where the process goes in the opposite direction. Note however that in the adiabatic case with the dark matter particles with mass $m \approx 1 - 2$ keV the perturbation spectrum is also cut off at $10^6 M_\odot$. So it is very much like the isothermal case.

The theory of the formation of the structure of the Universe is being actively developed now. It is well established that the simple one-component model with flat spectrum does not agree with observations. More complicated, multicomponent models with a nontrivial spectrum of fluctuations are considered. Numerical investigations of the behavior of systems with large number of gravitationally interacting particles are being done. The progress is obstructed because there are too many candidates for hidden mass and no one of them is well established. The same can be said about the fluctuation spectrum. It will be later shown that simple inflationary models lead to adiabatic fluctuations with a flat spectrum. But there exist interesting models leading to isothermal fluctuations with a spectrum which posseses a specific scale. The formation of the large scale structure of the Universe is already understood qualitatively and only quantitive details need to be worked out.The coming years may bring the solution to this grand problem.

10.9 Fluctuations of microwave background radiation

The standard cosmological model predicts cosmic microwave radiation with a Planck spectrum.

It is interesting that Gamow who first predicted the existence of the relic radiation in 1946, rather accurately estimated its temperature to be 6 K although his considerations were based on wrong assumptions. The Hubble constant was thought to be $H = 564$ km/s/Mps and therefore the age of the Universe was too small to let heavy nuclei be formed in stars. Gamow assumed that heavy nuclei were formed during primordial nucleosynthesis and estimated the temperature necessary for production of heavy nuclei.

The prediction of relic radiation was confirmed in 1965 by discovery of isotropic radiation with Planck spectrum and temperature 3K. This discovery was accidentally made by Penzias and Wilson, when they tested a new radio receiver designed for cosmic communication on wavelength 7.3 cm.

Dicke and his group specially looked for cosmic microwave radiation on 3 cm wavelength. Their results appeared later than that of the first group.

Earlier in 1964 A.G.Doroshkevich and I.D.Novikov calculated the spectrum of radiation of all the sources including the background radiation with different temperatures. They concluded that the microwave background should be easiest to find on 1 cm wavelength since the contribution of all other sources is the smallest here.

It is interesting that in 1940 an anomalous excitation of CN molecules in interstellar gas was found. The energy of the excited state is higher than that of the ground state by the value that is equal to the energy of a photon with $\lambda = 0.264$ cm and the degree of excitation corresponded to 3K radiation. It can be shown that such excitation proceeded only because of interaction with isotropic background radiation. The spectrum of the cosmic microwave background has been measured in many frequency bands and no deviations from a Planck spectrum were found.

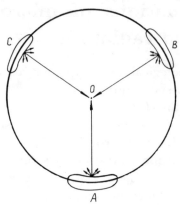

Figure 10.8: Three different regions A, B, C on the horizon of an observer in point O.

At high temperatures all particles, including photons were present in the primordial plasma. These photons are observed now as the cosmic microwave background. The thermal equilibrium of photons with other particles was broken at $z_{rec} = 1000$. The radiation comes to us from this epoch practically without any interaction with matter. We must stress that the expansion of the Universe does not alter the shape of Planck spectrum, but only the temperature.

The inflation leads to a high degree of homogeneity of matter density. This is also valid for photons. This conclusion is well confirmed by observations. The homogeneous distribution of photons in the early Universe leads to high isotropy of cosmic microwave background now. Figure 10.8 illustrates it. Different regions A, B, and C lie at the same distance from the observer O. The distance is the present horizon $r = 3H^{-1}/2$. Photons coming to the observer carry information about physical conditions in regions A, B, and C. The observer sees the three regions in different directions. Thus the homogeneity of plasma properties implies the isotropy of background radiation. This is one of the arguments in favour of inflation.

Discussing properties of horizon in the Friedman space in chapter 1 we made a conclusion, that regions which are farther away than H^{-1} could never interact. Thus until the moment of recombination,

t_{rec}, the size of causally connected regions is about ct_{rec}. Inside the regions of that size the temperature should be constant. However different regions could have different properties since they never interacted. In particular, their temperatures could be different. The angular scale of the cosmic microwave background anisotropy can easily be calculated. The angle at which a causally connected region at t_{rec} can be seen, is now:

$$\theta = t_{rec}H_0\frac{(1+z_{rec})^2}{2}(z_{rec}+1-\sqrt{z_{rec}+1})^{-1} \approx 2°$$

No such anisotropy has been observed. This is interpreted as a consequence of inflation.

Despite the inflation there should be some temperature fluctuations in the cosmic microwave background $\delta T/T$. There are four mechanisms leading to such fluctuations. The main effect on large scales is the Sachs-Wolfe effect. Briefly this is a shift of photon frequency in gravitational field.

The frequency ν of a photon passing through the potential difference $\Delta\phi$ changes as:

$$\frac{\delta\nu}{\nu} \approx \Delta\phi$$

In the case of a system of photons with Planck spectrum this results in the temperature change :

$$\frac{\delta T}{T} = \frac{\delta\nu}{\nu}$$

Thus, the photonic temperature in an inhomogeneous and nonstationary gravitational field depends on direction.

On smaller scales the main contribution to the temperature variation is given by Silk effect. It appears due to adiabatic density fluctuations. If entropy (the ratio of baryons to photons) is constant, baryon number density fluctuations lead to photon number density fluctuations. From the condition:

$$n_b/T^3 = const$$

it follows that:

$$\frac{\delta T(r)}{T} = \frac{1}{3}\frac{\delta\rho(r)}{\rho}$$

Inhomogeneities in the temperature distribution during recombination lead to an anisotropy in $\delta T/T$ now.

Along with the density fluctuation the Doppler effect connected with the peculiar motion of matter is also of importance. Peculiar motion is a random motion imposed on the Hubble expansion. The growth of fluctuations is possible only in the presence of peculiar motion of some kind. The distribution of peculiar velocities of galaxies at different z would be an excellent source of information about the primary fluctuations.

Another effect which also leads to temperature variation is the Zeldovich–Sunyaev effect (1982). It is not directly connected with perturbations of the metric at the recombination moment but is essential for the determination of the evolution of the Universe at $z = 5 - 10$. This effect originates from the energy transition by the hot electron gas to relic photons. The inhomogeneities in electron distribution in the sky lead to angular variation of the cosmic microwave background temperature.

This effect allows one to distinguish between the two theories of formation of large scale structure of the Universe, the pancake theory and the exploding instability theory. The latter explains the bubble structure of the Universe by entropy fluctuations. The scenario is as follows. Protostars with enormous mass up to $m = 10^2 M_\odot$ are formed due to entropy fluctuations. They give rise to shock waves of density when they explode. These waves initiate more explosions and this leads to the grid structure of the Universe. The calculations by Hogan (1984) show that because of the Zeldovich-Sunyaev effect this leads to big temperature fluctuations. Therefore the pancake theory is preferable.

The last mechanism leading to temperature anisotropy is the Doppler effect. An electron which absorbs and emits a photon changes its frequency according to the relation $\delta\nu/\nu \approx v$ where v is the deviation from the Hubble motion of the electron. The temperature in the direction of a moving cloud of electron gas is altered:

$$\frac{\delta T}{T} \sim v\,\tau_T$$

where v is the peculiar velocity of the cloud (the total velocity measured from the redshift is $H_0 R + v$) and τ_T is the optical depth of the cloud.

It is essential to find $\delta T/T$ during recombination, in order to obtain data on the early Universe. At that period $\delta T/T$ bears information about density fluctuations and gravitational waves formed during inflation (see chapter 11).

Only two mechanisms of $\delta T/T$ formation are essential at that moment: the Sachs-Wolfe and the Silk effects. They will be discussed in more detail.

The isotropic and homogeneous Friedman metric with small deviations can be written as follows:

$$ds^2 = a(\eta)^2(\eta_{\alpha\beta} + h_{\alpha\beta})dx^\alpha dx^\beta$$

Each volume of primordial plasma flashes for a short time Δt, when the optical depth of the plasma changes from $\tau \approx 1$ to $\tau \approx 0.1$. Let us suppose this happened at time η_e. The time of observation can be expressed through η_e as $\eta_0 = \sqrt{1 + z_{rec}}\,\eta_e \approx 30\,\eta_e$. The temperature we observe depends on the temperature at the moment of radiation:

$$T_0 = T_e\left(\frac{\eta_e}{\eta_0}\right)^2\left(1 + \frac{\delta T}{T}\right)$$

where

$$\frac{\delta T}{T} = \frac{1}{2}\int\left(\frac{dh_{ij}}{d\eta}e^i e^j - 2\frac{\partial h_{0i}}{\partial \eta}e^i\right)d\lambda \qquad (10.16)$$

and the integration is done from η_e to η_0. The vector $e^i = (\sin\theta\cos\phi, \sin\theta\sin\phi, \cos\theta)$ defines the direction of observation.

Equation (10.16) is the basic one in calculation of the Sachs-Wolfe effect. The increasing mode of density fluctuation can be written as follows:

$$\frac{\delta\rho}{\rho} = \frac{1}{2}h(k\eta)^2 e^{i(kx+\xi)} \qquad (10.17)$$

(compare it with eq.(10.5)). It is assumed that h depends on wave vector k and that $k\eta_{rec} < 1$. The relation between $\delta\rho$ and k is different for $k\eta_{rec} > 1$. Such values of k are not of interest, however, because fluctuations with the wavelength smaller than the horizon i.e. $k\eta < 1$ are quickly damped.

Substituting eq(10.7) into eq.(10.16) one finds for $\delta T/T$ (Grishchuk and Zeldovich, 1977):

$$\frac{\delta T}{T} = h e^{i(\xi + \frac{\pi}{2})} \Big[k\eta_0 \cos\theta - k\eta_e \cos\theta e^{ik(\eta_0 - \eta_e)\cos\theta}$$

$$+ 1 - e^{ik(\eta_0 - \eta_e)\cos\theta} \Big]$$

The relation between the temperature and the density fluctuations is very simple in the long wave limit $k\eta_0 \ll 1$:

$$\left(\frac{\delta T}{T} \right)_0 = \left(\frac{\delta\rho}{\rho} \right)_0 \cos^2\theta$$

where θ is the angle between the direction of observation and wave vector of the density waves. The relation becomes more complicated for smaller wavelengths.

The angular spectrum can be written down as the series expansion:

$$\frac{\delta T(\theta, \phi)}{T} = \sum_{n,m} C_{n,n} Y_n^m(\theta, \phi)$$

The spectrum of fluctuations $h(k)$ uniquely determines the angular distribution of $\delta T/T$. This relation, in the case of the flat spectrum $h(k) = Ak^{-3/2}$ (Starobinsky 1983), is:

$$\langle C_{nm}^2 \rangle = \frac{A^2}{100\pi n(n+1)}$$

The amplitude of the spectrum A determines density fluctuations according to the relation: $\delta\rho/\rho = Ak^2\eta^2/(20k^{3/2})$, and the moment when the primary structures were formed, which is defined by the condition $\delta\rho \approx \rho$.

Gravitational waves also lead to fluctuations of background radiation temperature. A flat monochromatic gravitational wave disturbs the metric as follows:

$$h_{\mu\nu} = -\frac{1}{3} B_{\mu\nu} k^3 \left[1 - \frac{(k\eta)^2}{10} + ... \right] e^{i(kr+\xi)}$$

where the traceless matrix $B_{\mu\nu}$ describes the tensor structure of the wave. The fluctuations of temperature of relic photons at $k\eta_0 < 1$ are (Grishchuk and Zeldovich, 1977):

$$\frac{\delta T}{T} = \frac{k^3}{60} \sin^2 \theta (C_1 \cos 2\phi + C_2 \sin 2\phi)(k\eta_0)^2$$

$$\times \left[\left(1 - \frac{\eta_e^2}{\eta_0^2} \right) \cos \xi - k\eta_e \cos \theta \left(1 - \frac{\eta_e}{\eta_0} \right)^2 \right]$$

where C_1 and C_2 are amplitudes of gravitational waves with different independent polarizations.

Gravitational waves lead to anisotropy of relic radiation in the same way as density fluctuations do. These two mechanisms can be distinguished by measurement of the polarization of relic radiation (Basko and Polnarev 1980, Sazhin 1984, Polnarev 1985).

Dipole anisotropy of the cosmic microwave background has been established (Smoot 1977). It is interpreted as a result of motion of the Galaxy and of the Sun system relative to the background. The angular scale of this anisotropy is large. It is about 180°. No anisotropy on smaller scales has been found. Every time an upper bound on fluctuations has been obtained. Up to now the strongest limits were obtained by "RATAN-600" and by the Moscow group in the "Relict" experiment. The quadrupole anisotropy is bounded by:

$$(\delta T/T)_Q < 3 \times 10^{-5}$$

The high degree of isotropy together with the developed large scale structure of the Universe seems to be an argument confirming the existence of dark matter.

Chapter 11

Quantum theory and the origin of fluctuations

The observed large scale structure of the Universe was formed out of small primary fluctuations. Until recently their origin presented a serious challenge to physicists. It was not clear what physical processes could produce fluctuations on large, cosmological scales. The inflationary model played an important role in the solution of this problem. Quantum fluctuations which usually have microscopic wavelength, become of very large wavelength in the inflationary universe because this wavelength increases exponentially, $\lambda \to \lambda \exp(Ht)$, and they become cosmologically significant. The amplitude of fluctuations also increases during inflation, but of course by a much smaller factor. Therefore galaxies, their clusters and other large scale structures, can be regarded as microscopic fluctuations blown up to macroscopic scales.

11.1 Fluctuations of scalar field

As it has been noted in chapter 6, the Universe expanded exponentially, while the scalar field $\phi_0(t)$ slowly approached equilibrium. The equation of motion of a scalar field in the de Sitter metric is the following:

$$(\partial_t^2 + 3H\partial_t)\phi - e^{-2Ht}\partial_i^2\phi = -\frac{\partial V}{\partial\phi} \qquad (11.1)$$

where $V(\phi)$ is the field potential ($V(\phi) = m^2\phi^2 - \lambda\phi^4$ in the simplest case) and the scale factor is $a(t) = a_0\exp(Ht)$. $\phi_0(t)$ is a space independent solution of equation (11.1).

It is well known that quantum fluctuations $\delta\phi(r, t)$ are imposed on the classical solution $\phi_0(t)$. It will be shown that quantum fluctuations alter the time, needed for $\phi_0(t)$ to reach equilibrium by:

$$\delta t(r) = -\frac{\delta\phi(r, t)}{\dot\phi_0(t)}. \qquad (11.2)$$

Indeed $\delta\phi(r, t)$ satisfies the equation:

$$\delta\ddot\phi + 3H\delta\dot\phi - e^{-2Ht}\partial_i^2\delta\phi = -\frac{\partial^2 V}{\partial\phi^2}\delta\phi. \qquad (11.3)$$

The term with spatial derivatives becomes negligible for large t and $\delta\phi$ starts to satisfy the same equation as $\dot\phi_0$. Therefore, the ratio between $\dot\phi_0$ and $\delta\phi$ approaches a constant (11.2), when $t \to \infty$. Thus, in the first order in δt one can write:

$$\phi(r, t) = \phi_0(t) + \delta\phi(r, t) \approx \phi_0(t - \delta t(r)) \qquad (11.4)$$

Hence, the transition from the de Sitter to the Friedman expansion in different regions of space starts at different times. This leads to fluctuations of metric:

$$h \approx H\delta t(r). \qquad (11.5)$$

The problem is a little more complicated because of the freedom in choice of time coordinate in the de Sitter space. The reader is asked either believe this order of magnitude estimate or to turn to original papers which we cite in what follows. The physical reason for the choice of a specific point on the time axis is the change of the expansion law.

In accordance with the general theory, metric fluctuations (11.5) lead to relative density fluctuations of the same value, when their wave length is equal to the horizon length (see chapter 10). Let us recall that δt is related to field fluctuations by the formula (11.2). Quantum fluctuations in the de Sitter space (strictly speaking, the difference between them and fluctuations in flat spacetime) have amplitudes of the order of H. Hence, density fluctuations are estimated to be :

$$\frac{\delta \rho}{\rho} = \frac{H^2}{\dot{\phi}_0(t)}.$$ (11.6)

The value of $\dot{\phi}_0$ can be found from equation (11.1). It is known that $\phi_0(t)$ changes slowly so the second derivative $\ddot{\phi}$ can be neglected. For the model potential $V(\phi) = -\lambda \phi^4$, which is typical for many theories, one obtains:

$$\phi_0(t) = \left[\frac{3H}{2\lambda(t_0 - t)} \right]^{1/2},$$ (11.7)

where t_0 is the time when inflation ends. In more realistic models the potential $V(\phi)$ differs from the one used here at large values of $\phi_0(t)$, so $\phi_0(t)$ does not tend to infinity but approaches a constant value. However this approximation is sufficient for our purposes. Taking into account that a wave with wavelength equal to H^{-1} is stretched $e^{H\tau}$ times by the expansion and acquires the length $k_0^{-1} = H^{-1} e^{H\tau}$, one can estimate the time necessary for this as:

$$\tau = H^{-1} \ln(H/k_0).$$ (11.8)

Thus, the density fluctuations at the horizon scale are:

$$\left(\frac{\delta \rho}{\rho} \right)_H = \lambda^{1/2} \ln^{3/2}(H/k_0) \times const.$$ (11.9)

The constant factor which can be found by more accurate calculation is approximately 0.1. These fluctuations are pure quantum ones. They originated from the zero-mode oscillations (see section 4 in chapter 5).

A flat spectrum of adiabatic fluctuations has thus been obtained, but the fluctuations are too large with natural values of λ. Indeed to become cosmologically significant the wavelength must increase $10^{20} - 10^{25}$ times so $\ln(H/k) \approx 50$. Taking $(\delta\rho/\rho)_H < 10^{-4}$, according to the relic radiation isotropy, demands $\lambda < 10^{-12}$. No natural model for such small λ is known.

The above considerations are only qualitative. The reader who wishes to find a more rigorous treatment, is referred to refs. [44, 46, 67, 36, 57]. A collection of original papers on inflationary models can be found in the book [75].

Inflationary models give rise to fluctuations needed for large scale structure formation. Moreover, the inflation leads to the observed homogeneity of our Universe on the average. Considering the extremely large scales (larger than the present horizon), it can be shown that the inflationary scenario predicts large inhomogeneities (at least in the frameworks of the chaotic inflation model). Random field fluctuations over the de Sitter background lead to the increase of field with time [57,70]:

$$\langle \phi^2 \rangle = H^3 t/4\pi, \tag{11.10}$$

when $H \gg m$ where m is the mass of the field. This can be considered as the random walk of the field with step H. It is known that the length of the path in this process is proportional to $\sqrt{N}$ after N steps. Physically this increase is connected with the blowing up of the short wave fluctuations when more and more modes contribute to $\langle \phi^2 \rangle$.

A very unusual behavior of ϕ takes place in the model of chaotic inflation with $\phi \gg m_{Pl}$. Classically the field tends to roll down the potential but quantum fluctuations pull it up. The Hubble constant $H = (8\pi\lambda\phi^4/3m_{Pl}^2)^{1/2}$ increases simultaneously, which leads to growth of fluctuations and to further increase of ϕ. Therefore, there exist some regions in which the field goes up the potential and only in a very small part of the Universe $\phi \to 0$ [58]. The characteristic scale of the process is, however, too large $L \gg H_0 \approx t_u \approx 10^{10}$ to observe any consequences of this model.

11.2 Gravitational waves

Detection of relic gravitational waves would confirm the existence of a de Sitter period in the history of the Universe.

As it has already been mentioned, metric tensor fluctuations can be divided into scalar, vector and tensor parts. The tensor part corresponds to gravitational waves. The amplitude, as well as the wavelength of these waves, is increased by inflation. The possibility of detecting the gravitational waves is based on this mechanism.

The evolution of gravitational waves is similar to the evolution of the scalar field discussed in the previous section. Generation of gravitational waves in inflationary models will be discussed in what follows. All other fields can be considered in the same way.

The metric tensor describing small fluctuations in the homogeneous and isotropic background can be written as:

$$g_{\mu\nu} = g_{\mu\nu}^{(b)} + h_{\mu\nu}, \tag{11.11}$$

where $g_{\mu\nu}^{(b)}$ is the background metric tensor and $h_{\mu\nu}$ represents small fluctuations. The Einstein equations can also be divided into the background and the fluctuation part. The equations for the background are the ordinary Friedman equations. The equations for fluctuations are:

$$\psi_{\mu\nu,\alpha}^{\alpha} + 2R_{\alpha\mu\nu\beta}^{(b)}\psi^{\alpha\beta} + R_{\mu\alpha}^{(b)}\psi_{\nu}^{\alpha} + R_{\nu\alpha}^{(b)}\psi_{\mu}^{\alpha} = 0, \tag{11.12}$$

where $\psi_{\mu\nu} = h_{\mu\nu} - \frac{1}{2}g_{\mu\nu}^{(b)}h_{\alpha}^{\alpha}$ and the index (b) refers to the background metric.

It is convenient to use the conformal time η. In terms of η the metric can be written as:

$$ds^2 = a^2(\eta)(d\eta^2 - dx^2 - dy^2 - dz^2).$$

In this coordinate frame equation (11.12) becomes very simple:

$$\ddot{h}_{\mu\nu} + \frac{\dot{a}}{a}\dot{h}_{\mu\nu} - \triangle h_{\mu\nu} = 0. \tag{11.13}$$

We look for solutions of this equation in the form:

$$h_{\mu\nu}(\eta, r) = \frac{\sigma(\eta)}{a(\eta)} G_{\mu\nu} e^{i(nr)/a} \qquad (11.14)$$

where n is the wave vector, $\sigma(\eta)$ is the amplitude of this wave and $G_{\mu\nu}$ is a constant second rank tensor describing the polarization state of the wave. For the wave moving along the axis OX, it is equal to:

$$G_{\mu\nu} = \begin{pmatrix} 0 & 0 & 0 & 0 \\ 0 & 0 & 0 & 0 \\ 0 & 0 & G_1 & G_2 \\ 0 & 0 & G_2 & -G_1 \end{pmatrix},$$

The scale factor has been inserted into expression (11.14) in order to explicitly take into account the adiabatic damping due to the expansion of the Universe.

The amplitude $\sigma(\eta)$ satisfies the equation of a parametrically excited oscillator:

$$\ddot{\sigma} + \left(n^2 - \frac{\ddot{a}}{a} \right) \sigma = 0. \qquad (11.15)$$

The energy is pumped to this oscillator by gravitational forces. Let us first consider the simple case when $\ddot{a} = 0$, i.e. $a(\eta) = a_0 \eta$. This corresponds to the radiation dominated Universe when the equation of state of matter is $p = \epsilon/3$. The solution of equation (11.15) in this case is a sum of two exponents:

$$\sigma(\eta) = A_1 e^{in\eta} + A_2 e^{-in\eta} \qquad (11.16)$$

A_1 and A_2 are the amplitudes of the waves going in direction $+n$ and $-n$, respectively.

This is the only case when the wave is not amplified. For all other equations of state gravitational waves are amplified (Grishchuk, 1974). A monochromatic wave can either be damped or amplified depending on the phase, but a random set of waves is always amplified in the average.

When $a(\eta)$ is a power function of η then $\ddot{a}/a \sim \eta^{-2}$. The expression in brackets in equation (11.15), corresponding to the square of frequency, becomes negative when $n\eta \ll 1$. Imaginary frequency means that oscillations increase.

Equation (11.15) cannot be solved analytically for an arbitrary function $a(\eta)$. We shall consider waves which are always longer than the horizon in order to understand the physical meaning of the equation. This condition is:

$$n^2 \ll \ddot{a}/a.$$

One solution of the equation can be easily guessed, $\sigma(\eta) = B_1 a(\eta)$. The second one is obtained in the standard way. Thus, the general solution is:

$$\sigma(\eta) = B_1 a(\eta) + B_2 a(\eta) \int \frac{d\eta}{a^2(\eta)}. \qquad (11.17)$$

The first term in eq. (11.17) increases with the scale factor and the second one decreases. The initial condition for $\sigma(\eta)$ is naturally to be fixed at the radiation dominated era when both modes are well determined [1]. They are expressed through the amplitude of quantum oscillations $h\omega/2$. After the period of amplification when the solution is given by expression (11.17), the radiation domination era again takes place. Equation (11.16) describes metric fluctuations again, but the coefficients are different:

$$\sigma(\eta) = C_1 e^{in\eta} + C_2 e^{-in\eta}. \qquad (11.16)$$

The metric fluctuations are matched at the moment of the transition from one expansion mode to another, so that σ and its first derivative are continuous. This condition allows one to express the coefficients C_1, C_2 through B_1, B_2 and finally through A_1, A_2. The relation between C and A is:

$$C_1 = \frac{H}{in} Z^2 (A_1 + A_2) + \frac{1}{2} Z^2 (A_1 - A_2)$$

$$C_2 = -\frac{H}{in} Z^2 (A_1 + A_2) + \frac{1}{2} Z^2 (A_1 - A_2)$$

---

[1] This era could have taken place before inflation. However, the exact form of the expansion law before inflation is not important.

where H is the Hubble parameter during inflation with energy density equal to the false vacuum energy ρ_V; $H^2 = 8\pi\rho_V/3m_{Pl}^2$ and Z is the ratio between scale factors at the beginning and at the end of inflation:

$$Z = \exp\{H(t_f - t_s)\}.$$

The mean square amplification, after averaging over amplitudes, is:

$$\left(\langle|C_1|^2\rangle + \langle|C_2|^2\rangle\right) \approx \frac{H^2}{n^2}Z^4\left(\langle|A_1|^2\rangle + \langle|A_2|\rangle^2\right).$$

The final result for the energy density of the waves with frequency 10^{10}Hz $> \nu > 10^{-14}$Hz is:

$$\rho_g(\nu) = A\rho_\gamma \frac{\rho_v}{m_{Pl}^4}\frac{1}{\nu}, \tag{11.18}$$

where A is a coefficient of the order of unity, and ρ_γ is the relic radiation energy density. For the longer waves one obtains:

$$\rho_g(\nu) = A\rho_c\frac{\rho_v}{m_{Pl}^4}\left(\frac{H_0}{\nu}\right)\frac{1}{\nu}, \tag{11.19}$$

where $\rho_c = 3H_0^2/8\pi G$ is the critical energy density, and H_0 is the contemporary value of the Hubble constant [63].

Gravitational waves generated during the de Sitter period have an extremely wide frequency spectrum: 10^{10}Hz $> \nu > 10^{-17}$Hz. The waves with wavelength about the horizon size have the strongest impact on the electromagnetic background. Relic photons propagate in a variable gravitational field of these waves and change their frequency. This leads to angular variation of the microwave radiation temperature.

No anisotropy of cosmic microwave background has been found yet. There are only upper bounds on it. Hence, one can obtain the following bound on the value of the Hubble parameter during inflation. It is, in terms of Planck mass:

$$H_{inf} \leq 10^{-4}m_{Pl}. \tag{11.20}$$

11.3 Isothermal fluctuations

Density fluctuations in inflationary models were discussed in section 1. These are adiabatic fluctuations. The generation of baryon asymmetry, described in chapter 8, leads to adiabatic fluctuations too, since the value $(N_B - N_{\bar{B}})/N_\gamma$ is determined by fundamental constants and is homogeneous. Thus the adiabatic fluctuations became very popular all the more since the structure of the Universe can be qualitatively explained by them, while the isothermal fluctuations were considered as an exotic example. The development of the theory showed, however, that the isothermal fluctuations may be as natural as adiabatic ones. In what follows we briefly discuss some models with adiabatic fluctuations.

In one of the models considered by Linde, adiabatic fluctuations are connected with the energy density fluctuations in the hidden sector of the theory. The hidden sector contains particles which interact with usual matter only gravitationally. In the modern zoo of elementary particles predicted by different theories one can easily find a number of candidates for that. Fluctuations of energy density of such particles do not alter the plasma temperature and so are isothermal. One can get sufficiently large isothermal fluctuations in a simple model with two scalar fields ϕ_1 and ϕ_2. It is assumed that the field ϕ_1 belongs to the hidden sector and has a larger self-interaction coupling constant. Big isothermal fluctuations can be obtained if the energy density of this field decreases slower than the energy density of all other fields. In this model the ratio N_B/N_γ and the temperature are constant at an early stage, while the fluctuations of the energy density in the hidden sector become significant. The term isothermal fluctuations are, however, usually applied to the case when the ratio N_B/N_γ changes.

In paper [41] the mechanism of generation of large fluctuations of N_B/N_γ by small fluctuations of temperature was considered. In the model of this work the baryosynthesis proceeds at small temperatures. The value of baryonic excess depends exponentially on temperature after a phase transition in this model. Hence, small temperature fluctuations alter the number of baryons by several orders of magnitude.

The examples considered here prove that not only adiabatic but also isothermal fluctuations might have played a role in the formation of the structure of the Universe. One should not confine oneself to a single possibility but should keep one's mind open for other variants which might help to solve the problem of the structure formation with very small fluctuations of temperature of the cosmic microwave background.

11.4   Strings and initial fluctuations

The growth of quantum field fluctuations during inflation is not the only possibility to generate enegy density fluctuations. New theories predicting density fluctuations due to cosmic strings [72,69] were developed in recent years. Strings are very long and thin configurations of scalar field. Strings as well as monopoles and domain walls are stable topological defects which might have been formed during phase transitions in the early Universe. The following example explains how they appear. Let us consider a scalar field with the well-known potential:

$$V(\phi) = \lambda(\phi^*\phi - \eta^2)^2.$$

The stable state of the field is:

$$\langle\phi\rangle = \eta e^{i\Phi}, \tag{11.21}$$

with arbitrary constant phase Φ. This phase is determined by random fluctuations of the field ϕ during the phase transition from $\langle\phi\rangle = 0$ to $|\langle\phi\rangle| = \eta$. Let L be the correlation length of the fluctuations (L must be smaller than the horizon). Phases over distances bigger than L are not correlated. The change of the phase along a closed curve does not have to be zero but it may be:

$$\Delta\Phi = 2\pi n, \quad n = 0, 1, 2\dots. \tag{11.22}$$

There must be a point where $\phi = 0$ inside a curve with e.g. $n = 1$. So the condition $|\langle\phi\rangle| = \eta$ is not fulfilled everywhere in the space. This results in the formation of a nontrivial configuration of field ϕ with nonzero energy, i.e. of a string. The strings should either be infinite or closed. It can be shown that the transverse size of a string is $R \sim \eta^{-1}$ and its linear mass density $\mu \sim \eta^2$ up to the factors depending on the coupling constants.

Closed strings, smaller than the horizon, quickly disappear because of their tension. Only the strings with characteristic curvature radius exceeding the horizon $L_h \approx t$ are interesting for cosmology. It can be shown that there must be several such strings inside the volume bounded by L_h and that their energy density is by the order of magnitude equal to :

$$\rho_s \approx \frac{\mu t}{t^3} = \frac{\mu}{t^2}. \tag{11.23}$$

Density fluctuations of matter should be of the same order, and so the relative density contrast for fluctuations with wavelength L_h is:

$$\frac{\delta \rho}{\rho} = \frac{\mu/t^2}{(3m_{Pl}^2/32\pi t^2)} = \frac{32\pi}{3} \frac{\mu}{m_{Pl}^2}. \tag{11.24}$$

The numerical coefficient in this formula is correct in the radiation dominated Universe. For the matter dominated Universe the factor $32\pi/3$ must be changed to 6π.

Therefore, strings as well as scalar fields, lead to scale independent density fluctuations. Such fluctuations were discussed in chapter 10. The value of $\delta \rho/\rho$ depends on μ. It is necessary that $\mu = (10^{15} \text{GeV})^2 - (10^{16} \text{GeV})^2$ in order to obtain the density fluctuations needed to form galaxies i.e. $\delta \rho/\rho = 10^{-4} - 10^{-5}$. This happens to be the energy scale of the grand unification theories.

The reader may find a detailed discussion of strings and how they can be observed, in the review article by A.Vilenkin [69].

A shortcoming of the string model is that cosmic strings may exist but they do not have to, in contrast to magnetic monopoles which are unambiguously predicted by Grand Unification models.

The reader can see that at the moment there exist no definite models for the origin of the energy density fluctuations. A few years ago not a single mechanism was known, and the origin of the density fluctuations in the Universe was mysterious. We hope that the analysis of predictions of different models will allow one to make a choice between possibilities and to formulate at last a complete theory of the formation of the large scale structure of the Universe.

References

[1] N.N.Bogoliubov and D.V.Schirkov, *Vviedienye v teoryu kvanto-vannyh polyey*, M.Nauka (1976) (in Russian). Introduction to quantum field theory.

[2] W.B.Bragintsky and W.I.Panov, *ZETF* (1971), **61**, 873.

[3] M.I.Vysotsky, A.D.Dolgov and Ya.B.Zeldovich, *Pisma v ZETF* (1977), **26**, 200.

[4] S.Weinberg, *Gravitation and Cosmology*.

[5] S.Weinberg, *The First Three Minutes*.

[6] S.S.Gerschteyn and Ya.B.Zeldovitch, *Pisma v ZETF* (1966), **4**, 174.

[7] W.L.Ginzburg, D.A.Kirzhnits and A.A.Liuboschin, *ZETF* (1971), **60**, 451.

[8] L.P.Grishchuk, *ZETF* (1974), **67**, 825.

[9] V.T.Gurovich and A.A.Starobinsky, *ZETF* (1979), **77**, 1699.

[10] A.D.Dolgov, *ZETF* (1980), **79**, 337.

[11] A.G.Doroshkievich and I.D.Novikov, *DAN SSSR* (1964), **154**, 809.

[12] Ya.B.Zeldovich, *Pisma v ZETF* (1976), **24**, 29.

[13] Ya.B.Zeldovich, L.P.Grishchuk, *Uspekhi Fizicheskyh Nauk* (1986), **149**, 695.

[14] Ya.B.Zeldovich and I.D.Novikov, *Stroienie i evoliutsya Vseli-ennoy*, M.Nauka (1975) (in Russian). The Structure and Evolution

of the Universe.

[15] Ya.B.Zeldovich, L.B.Okun and S.B.Pikelner, *Uspekhi Fizich-eskyh Nauk* (1965), **87**, 115.

[16] Ya.B.Zeldovich and D.D.Sokolov, *Uspekhi Fizicheskyh Nauk* (1985), **146**, 493.

[17] Ya.B.Zeldovich and R.A.Sunyaev, *Astrofizika i kosmicheskaya fizika*, M.Nauka (1982) 9.

[18] A.A.Klyupin, M.V.Sazhin, I.A.Strukov and D.P.Skulachev, *Pisma v Astronomichesky Zhurnal* (1986) No.5.

[19] D.A.Kirzhnits, *Pisma v ZETF* (1972), **15**, 745.

[20] L.D,Landau and E.M.Lifshitz, *Teoria polya*, M.Nauka (1973) (in Russian). Field Theory.

[21] A.A.Logunov, *Novye predstavlenya o prostranstve, vremeni i gravitatsyi*, M, MGU (1986).

[22] E.M.Lifshitz, *ZETF* (1946), **16**, 587.

[23] V.A.Liubimov, E.G.Novikov, V.Z.Nozik, E.F.Tretyakov, V.O.Kozik and N.F.Myasoedov, *ZETF* (1981), **81**, 1959.

[24] S.P.Mikheev and A.Yu.Smirnov, *Yadernaya fizika* (1985), **42**, 1441.

[25] L.B.Okun, *Leptony i kvarki*, M.Nauka (1981) (in Russian). Leptons and Quarks.

[26] L.B.Okun, $\alpha, \beta, \gamma...Z$, M.Nauka (1982).

[27] L.B.Okun, *Fizika elementarnykh czastic*, M.Nauka (1984) (in Russian). Elementary Particle Physics.

[28] P.J.E.Peebles, Large Scale Structure of the Universe.

[29] A.V.Rubakov, *Pisma v ZETF* (1981), **33**, 658; *Nucl.Phys.* (1982), **B203**, 311.

[30] M.V.Sazhin, *Sovremennyje teoreticzeskije i eksperimentalnyje problemy teorii otnositelnosti i gravitacji*, M.LGPI im. Lenina (1984) 88.

REFERENCES   239

[31] A.D.Sakharov, *Pisma v ZETF* (1967), **5**, 32.

[32] A.A.Starobinsky, *Pisma v ZETF* (1979), **30**, 719.

[33] L.D.Fadeev, *UFN* (1982), **136**, 435.

[34] F.V.Schwartzman, *Pisma v ZETF* (1969), **9**, 315.

[35] A.Albrecht and P.Steinhardt, *Phys. Rev. Lett.* (1982), **48**, 122.

[36] J.Bardeen, P.Steinhardt and M.Turner, *Phys. Rev.* (1983), **D28**, 679.

[37] A.Belavin, A.Polyakov, A.Schwartz and Yu.Tupkin, *Phys. Lett.* (1975), **B59**, 85.

[38] C.G.Callan Jr., *Phys. Rev.* (1982), **D25**, 2141; **D26**, 2058.

[39] A.D.Dolgov, preprint ITEP – 13.1987, *Einshteinovsky Sbornik*, M. Nauka (1986)(in Russian).

[40] A.D.Dolgov, A.D.Linde, *Phys. Lett.* (1982), **116B**, 329.

[41] M.Fukugita and V.A.Rubakov, *Phys. Rev. Lett.* (1986), **56**, 988.

[42] G.Gamow, *Phys. Rev.* (1946), **70**, 572.

[43] A.Guth, *Phys. Rev.* (1981), **D23**, 347.

[44] A.Guth, S.-Y. Pi, *Phys. Rev. Lett.* (1982), **49**, 1110.

[45] S.Hawking, *Nature* (1974), **248**, 331.

[46] S.Hawking, *Phys. Lett.* (1982), **115B**,295.

[47] C.Hogan, *Ap.J.* (1984), **284**, L1.

[48] E.P.Hubble, *Proc. Nat. Acad. Sci. USA* (1929), **15**, 168.

[49] P.Hut, *Phys. Lett.* (1977), **69B**, 85.

[50] J.Jeans, *Phil. Trans.* (1902), **199A**, 49.

[51] D.A.Kirzhnits and A.D.Linde, *Phys. Lett.* (1972), **42B**, 471.

[52] V.A.Kuzmin, V.A.Rubakov and M.E.Shaposhnikov, *Phys. Rev. Lett.* (1985), **155B**, 36.

[53] T.D.Lee and C.N.Yang, *Phys. Rev.* (1955), **98**, 1501.

[54] W.Lee and S.Weinberg, *Phys.Rev.Lett.* (1977), **39**, 165.

[55] A.D.Linde, *Phys. Lett.* (1983), **129B**, 177.

[56] A.D.Linde, *Phys. Lett.* (1983), **108B**, 389.

[57] A.D.Linde, *Phys. Lett.* (1982), **116B**, 335.

[58] A.D.Linde, *Phys. Lett.* (1986), **175B**, 395.

[59] B.B.Mandelbrot, *Fractals, form, chance and dimension,* San Francisco (1977).

[60] M.A,Markov, Very early universe, *Proc. of the Nuffield Workshop,* Eds. G.Gibbons, S.Hawking, S.T.C.Siklos, Cambridge (1982) 353.

[61] A.McKellar, *Publ. Astron. Soc. Pacific.* (1940), **52**, 187.

[62] A.A.Penzias and R.W.Wilson, *Astron.J.* (1965), **142**, 419.

[63] V.A.Rubakov, M.V.Sazhin and A.V.Verjaskin, *Phys. Lett.* (1982), **115B**, 189.

[64] K.Sato and H.Kobayashi, *Progr. Theor. Phys.* (1977), **58**, 1775.

[65] G.F.Smoot, M.V.Gorenstein and R.A.Muller, *Phys. Rev. Lett.* (1977), **39**, 898.

[66] A.A.Starobinsky, *Phys. Lett.* (1980), **91B**, 99.

[67] A.A.Starobinsky, *Phys. Lett.* (1982), **117B**, 175.

[68] R.A.Sunyaev and Ya.B.Zeldovich, *Astrophys. and Space Sci.* (1969), **4**, 301.

[69] A.Vilenkin, *Phys. Reports* (1985), **121**, 263.

[70] A.Vilenkin, L.H.Ford, *Phys. Rev.* (1982), **D26**, 1231.

[71] H.Yukawa, *Proc. Phys.-Math. Soc. Japan* (1935), **17**, 48.

[72] Ya.B.Zeldovich, *Mon. Not. Roy. Astr. Soc.* (1980), **192**, 663.

[73] Ya.B.Zeldovich, *Sov. Sci. Rev. E. Astrphys. and Space Phys.* (1984), **3**, 1.

[74] Ya.B.Zeldovich and L.P.Pitajevsky, *Commun. Math. Phys.*

(1971), **23**, 185.

[75] *Inflationary Cosmology*, ed. L.F.Abbott and So-Young Pi, World Scien., Singapore, 1986.

[76] Ya.B.Zeldovich and M.Yu.Khlopov, *Phys. Lett.* (1978), **79B**, 239.

[77] V.A.Belinsky, L.P.Grishchuk, Ya.B.Zeldovich and I.M.Khalatnikov, *Phys. Lett.* (1985), **155B**, 232; *ZETF*, **89**, 376.

[78] A.G.Polnarev, *Astronomichesky Zhurnal* (1985), **62**, 1041.

[79] M.M.Basko and A.G.Polnarev, *Astronomichesky Zhurnal* (1980), **57**, 465.

Index

Basics of
Modern Cosmology

"Basics of"

a series edited by
J. Tran Thanh Van

Basics of Electron-positron Collisions
F. M. Renard

Basics of Lie Groups
M. Gourdin

Basics of Cosmic Structures
L. M. Celnikier

Basics of Modern Cosmology
A. D. Dolgov, M. V. Sazhin, Ya. B. Zeldovich

Basics of Functional Methods and Eikonal Models
H. M. Fried

Cover illustration :   *Michelangelo*
"Division of the Light from the Darkness"
Monumenti Musei e Gallerie Pontificie
(Città del Vaticano)